最有分寸
溝通術

The Leader Phrase B⬤ok 3000+ Powerful Phrases That Put You In Command

派屈克·亞倫 Patrick Alain ——著

陳松筠 ——譯

▎謝詞

　　我最感謝的人就是采娜，我的妻子，在我籌備這本書期間，她無私的付出所有的愛與支持，一路陪伴著我。

　　也謝謝我的小女兒蜜雪兒，與我一起分享許多特別的時光。蜜雪兒，你是如此地美麗，徹底改變了我的世界。因為有妳，我開始學習做一名父親，再也沒有比這更棒的事。

　　我還要感謝湯姆‧卡洛和蜜雪兒‧拉娜，他們總是在我需要的時候伸出援手，我由衷感激他們的指導。

　　另外，我的編輯克絲汀‧達莉也功不可沒。她不僅投入全部心力，更讓整本書的編輯過程輕鬆愉快。

　　謝謝麗娜和巴山，沒有他們，我不會有今天的成績。

　　親愛的讀者，希望你能仔細讀完本書裡的一字一句。雖然不知道你的名字，但是我由衷地感謝你。

前言

　　任何人碰上預料之外的狀況都會感到棘手，特別是在上位的領導者。不管你是執行長、經理人、教練、律師、醫生、政治人物、業務、訓導主任，甚至是國家首長，大家都期待你能掌握所有資訊並且改善現況，帶領所有人邁向更美好的未來。這一點也不輕鬆。如果你又嘴拙，説話不懂得修飾，那壓力更是大。既然你選擇打開這本書，想必是對現狀不甚滿意。你知道你表現得可以更好，只是缺了能點石成金的詞句。而且因為你很忙碌，一套快速簡單的教材最符合你的需求，幫你成為不論內外都熠熠發亮的辦公室之星。

　　本書列出許多能讓你脫穎而出的應對語句，堪稱出版界創舉。筆者的目的是和大家分享這些談話亮點，讓你不管在任何場合都能展現出可靠、高信服力的領導特質。過去，想要更上一層樓的經理人會參加大大小小的研討會以獲取幫助他們升官的技巧，這一套已經過時了。現在你爬得越高，越要隨時準備好運用強而有力的語句來激勵人心和下達指令。最高境界是內化這些句子成為你生活的一部分，隨時隨地都能脫口而出。因此，本書分成八大部分，簡單易懂，最後還附贈紅利放送區（Bonus）。

　　不管你已經多麼成功，你的説話與聆聽技巧應該還可以更上一層樓。書中的技巧能夠加強你的語文能力，進而展現出你的自信與領導風範。只要你在每個場合都能説得鞭辟入裡、恰到好處，保證不管於公於私都會有意想不到的收穫。

　　現在讓我們一起開始這趟威力溝通之旅！

如何使用本書

　　本書的設計很有彈性。你可以從頭讀到尾，對所有內容先有個大致的概念；也可以針對比較有用的單一主題或溝通技巧深入研究，舉例來說，你可能對於在困頓時刻向大眾喊話很拿手，卻不擅長處理與其他人的衝突，那麼請多多善用目錄，尋找對你最為有用的章節。

　　看熟所有的例句之後，挑出最適合你自己使用的那些句子。然後，反覆練習直到能自然地脫口而出。一旦需要的時候，就能夠從容自信地運用這些話語。你也可以加入一點自己的用詞，畢竟每個人說話的習慣不盡相同。請記住，語調、肢體語言，和時機也會影響你的表達。有時候因為表達方式或談話對象不同，同一句話在這個場合顯得機智幽默，換到那個場合則顯得無禮失當。雖然本書所列的句子已經是針對特殊情況的通用句，但絕非一切都適用，請善用自己的判斷。尤其當你想用幽默的句子來強化效果，有時會顯得無禮或傲慢，而且不是每個人都能明白你話中有話。總之，請仔細推敲。

　　本書中每個情境都附有「態度量表」，可以當做小抄來使用，協助你選擇合適的情緒。舉例來說，當你遇到下頁的狀況時態度指標由「緩和」到「迎戰」，每個句子的態度將順著這個方向逐漸遞增。如果你想當和事佬把大事化小，多用點前面的句子；如果你想要直截了當的反擊，就選用靠近結尾的那幾句。

　　最後，筆者想強調，本書不只是單純列出所謂「合宜」的語句，也會有些口語和通俗用語。本書原文中的英文是一般美國人都了解並且使用的，在加拿大、澳洲等地區可能就不是這麼說。說到底，語言是活的，語言學也是日新月異。每天都有新的字或片語出現，

意義發生變化，甚至被揚棄不用。因此，筆者十分了解這本書的內容將會隨 著日常語言的演化而需要不定時的更新。想了解更新的內容，請上筆者的網站 www.patrickalain.com，也竭誠歡迎你的意見與指教。

※ 編按： 1. 由於本書原文為英文，因此所有例句皆以中英對照形式呈現。
　　　　 2. 原文英文語句語氣態度有明顯的差異，但部分語句翻成中文之後，語氣表達的程度順序可能會有些許差異，讀者可以視個人需求自行調整。

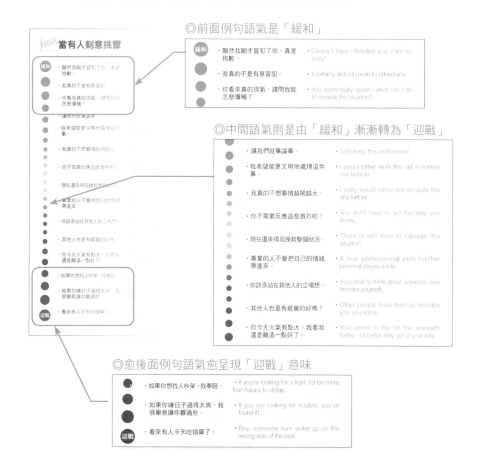

◎前面例句語氣是「緩和」

◎中間語氣則是由「緩和」漸漸轉為「迎戰」

◎愈後面例句語氣愈呈現「迎戰」意味

9

最有分寸溝通術
129 種情境，3000 句客氣話、機車話通通到位！

1 日常對話

2 辦公室厚黑學

3 處理衝突與憤怒
Part

4 Part 展現外交手腕

5 談判時刻

6 解決問題

表現風度禮貌

掌控發言權的帝王學

Bonus 領導者可以這樣說

Part

日常對話

如何在日常對話中，表達自己的立場，無論
是贊同、反對還是打探情報，使用適當的句
子更能有效達到溝通目的。

溝通哪些事

- 如何表示同意
- 如何表示反對
- 如何開啟對話
- 如何結束談話
- 如何傳遞消息
- 如何闡述觀點
- 如何詢問他人意見
- 如何表達自己的意見
- 如何尋求更多的資訊
- 如何澄清自己的觀點

- 如何向他人傾吐祕密
- 如何轉換話題
- 如何表達懷疑
- 如何和失聯已久的舊識重拾連絡
- 當你回答完一個問題之後
- 當你不想回答問題時
- 當別人請你重講一次

> **好的對話讓人銘記在心**
> ——約翰布朗（John Mason Brown，美國劇評家及作家）

　　說話是一門值得深究的技巧。交談的時候不只是要傳遞對的訊息，更要避免一些一般人常犯的錯誤。以下是六個讓你展現領袖風範的關鍵談話技巧：

1 簡潔扼要

　　想要成為領導者就要學會如何簡單清楚的表達意見。沒有人喜歡又臭又長或是不斷重複的對話。川普企業的董事長兼 CEO 唐納川普就是談話簡潔扼要的例子。他不會用一堆毫無意義的詞藻，反而是直接了當的說重點。在他的著作《如何致富（How to Get Rich）》中，川普寫道：「愛耍嘴皮子成就不了大事。」短短幾個字道出所有重點，說實在的，筆者真的想不出還有其他更好的說法。

2 增加詞彙

　　千萬不要以為自己已經知道夠多字彙了，每天盡量吸收一到兩個新詞。閱讀報章雜誌，畫下你沒看過的詞彙；學習如何正確使用，並且練習納入你的日常對話中。如果你比一般其他人多了一千個辭彙，自然容易脫穎而出，聽起來你就會像個領導者。

❸ 精確使用成語

微調你的語言,然後適時地展現出來。這是一個沒有盡頭的過程,不論何時何地,盡量使用生動靈活的通俗用語或成語,像是「美國時間」或是「大仁哥」。當然,請避免使用陳腔濫調,才能讓其他人更容易接受你說的話,也會留下更深的印象。

❹ 借他人之口

領導人在組織裡爬得越高,承受的風險也越大。與在收發室工作或當小業務的時候比起來,領導者每次開口都要謹慎,因為他們的發言可以激勵人心,卻也會讓人欲振乏力。領導人可能會成為晚間新聞的風雲人物,也可能變成社會頭條的貪官汙吏。能見度越高,發言的影響力越大。因此,很多企業的領導者或政府官員都會選擇使用公關或發言人。如果你可以,請盡量借他人之口。

❺ 他山之石可以攻錯

當你碰上了拙劣的溝通對象,先想想自己是不是也犯了一樣的錯誤。留心生活中好的溝通範例,久而久之自然會進步。

❻ 加強自己的論述

領導人一般會用生活軼事、時下新聞事件,或是歷史來強化自己的論點。如果想獲得最佳效果,可以在談話中插入最近的重大新聞事件、體育競賽結果、新上映的電影,或是自己的生活遭遇。

001 如何表示同意

明確

・沒錯！	・True!
・一點兒也沒錯。	・That's absolutely true.
・和我想的一樣。	・You're reading my mind.
・我不能同意你更多。	・I couldn't agree more.
・我打從心裡同意。	・I agree with you wholeheartedly.
・我同意。	・I agree.
・我們是同一陣線。	・We're on the same page about this.
・這個觀點很好。	・That's a good point.
・我完全同意。	・I'm in complete agreement.
・這樣就說得通了。	・That makes total sense.
・我完全明白你的意思。	・I know exactly what you mean.
・我倆看法一致。	・We see eye to eye.
・你是對的。	・You're correct.
・我的意見跟你一樣。	・My opinion corresponds with yours.
・我們想法雷同。	・Our opinions coincide very well.
・你說到我認為的重點了。	・You've touched on the essence of what I was trying to say.
・我們的想法完全相同。	・Our thoughts are in complete accord.
・我對你的意見心悅誠服。	・I'm in full agreement with you.
・我支持你的看法。	・I subscribe to your point of view.

- 我想大家都接受你的看法。
- I think your point is very well taken.

- 很高興我們的想法竟然不謀而合。
- I'm glad to see that we're on the same page.

- 我想你是對的。
- Common sense tells me you're right.

- 我站在你這邊。
- We're on the same wavelength.

- 我完全理解你的想法。
- I'm square with you on that.

- 我沒有異議。
- I have no problem with that.

- 你的意見我收到了。
- Your view of things is well received.

- 你的看法確實有些道理。
- Your point has relevance.

- 我不反對你的意見。
- I have no qualms with your perception of things.

- 我勉強同意。
- I concede the point.

- 我放棄,算你贏。
- I give up—you win.

- 既然你打算爭到贏,那就照你的意思吧!
- You're going to keep arguing until I give up, so have it your way.

含糊

- 反正你永遠是對的!(諷刺語氣)
- You're right, as always. [sarcasm]

002 如何表示反對

客氣

- 我們之間有些觀念上的歧異。
- We appear to have a divergence of beliefs.

- 我了解你的意思，但是…
- I understand the point you're making, but...

- 我完全明白你說的，但是…
- I certainly get what you're saying, but...

- 我尊重你的看法，但是…
- I respect your point of view, but...

- 我並不完全同意你的看法。
- I don't entirely agree with you on that.

- 我必須委婉地表達反對。
- I must courteously disagree with you.

- 我尊重你的意見，但不代表同意。
- I respectfully disagree.

- 我倒不覺得如此。
- That's not the way I see it.

- 這是其中一種看法，但並不正確。
- That's one way to look at it, but it's not the right way.

- 這個領域的專家應該都不會同意你。
- Few experts on this subject would agree with you.

- 現在的情況其實有幾種不同的可能。
- There's more than one way to look at this situation.

- 我聽到了很多流言蜚語。
- There's a lot of speculating going on here.

- 很明顯的，我們彼此詮釋的角度不同。
- There's obviously a divergence in interpretation.

- 我反對。
- I have a point of disagreement.

- 我想我們都同意雙方沒有什麼共識。
- I guess we're going to agree to disagree.

- 我對這點強烈質疑。
- I have great doubts about that.

- 我不同意，但是我重視你的意見。
- Not really, but I value your opinion.

- 還有其他更好的主意。
- There's a better way to look at this.

- 你的立論前提有瑕疵。
- Your premise is a bit flawed.

- 你應該再好好想一想，因為你的說法有問題。
- You might want to look it up because that's not right.

- 我沒有辦法同意你。
- You leave me no choice but to disagree with you.

- 我不認同你的看法。
- I don't subscribe to this point of view.

- 我只能持反對意見。
- I have no option but to disagree with you.

- 我們的意見根本南轅北轍。
- Our opinions are radically different.

- 我無法同意，因為你顯然是錯的。
- My disagreement with you stems from the simple fact that you're wrong.

- 我完全不同意。
- I disagree with you completely.

- 這根本是胡說八道。
- There is no truth to that whatsoever.

- 你根本不懂。
- You need more education on the matter.

- 你完全搞不清楚狀況。
- You're out of touch.

- 你真是錯得離譜。
- You couldn't be more wrong.

- 你已經離譜到根本不知道自己錯在哪裡。
- You're so wrong, you don't even know how wrong you are!

直接

25

003 如何開啟對話

禮貌

- 真高興能和你談話。
- It's great talking to you.

- 我很期待和你深入討論這個話題。
- I can't wait to get deeper into this topic with you.

- 我一直想和你好好聊一聊。
- I've wanted to talk with you for a long time.

- 我很希望能和你談一談。
- I would appreciate having a conversation with you.

- 我需要花點時間和你談一件事情。
- I'd like to talk to you about something for a moment.

- 你現在有空嗎？我希望和你討論一下。
- Do you have a minute? I would like to discuss something with you.

- 如果能知道你對…的看法，對我幫助很大。
- It would help me a lot to know your opinion on...

- 我真的很想和你討論一下這件事。
- I really want to have a dialogue with you about this.

- 我很認真在聽，也想了解你的意思。
- I'm listening to you and I want to understand.

- 今天，我希望能討論…
- Today, I'm hoping we can discuss...

- 請自在表達你的看法。
- Please feel free to speak openly.

- 不管今天討論的結果是什麼都好。
- Whatever results from our discussions will be just fine.

- 我希望我們的討論能獲得具體的結果。
- I'd like for our conversation to lead to something concrete.

- 我一向有話直說，你呢？
- I've always been good with conversing freely—how about you?

- 我想再多談談…
- I'd like to elaborate a bit on...

- 這個問題有點複雜，我們應該多花些時間聊聊。
- This is a tricky subject—let's talk about it for a while.

- 我們試著解決這件事吧！
- Let's take a whack at this, shall we?

- 事到如今，能討論的越深入越好。
- At this point, the more we talk, the better it will be.

- 有些話我不吐不快。
- I need to get my point across.

- 如果能按部就班談下去，應該會有所進展。
- We'll get further if we can get along from the start.

- 我希望能夠把事情搞得更清楚些。
- My goal is for this to become less ambiguous.

- 請多說點理由來說服我。
- Please help me get comfortable with your side of the argument.

- 何不開門見山地聊一聊，沒有壞處不是嗎？
- Why not simply talk about it? What could be the harm in that?

- 我不是只說說而已，是真的想做點事。
- I'm not looking to simply talk—I'm in this to make something happen.

- 我今天專程來聽聽你的意見，看有什麼可以幫上忙的地方。
- I'm seeing you today to hear you out and possibly help you.

- 我們早就該好好談談這件事了。
- We should have started this discussion long ago.

- 我也不想講這些，但真的沒其他辦法。
- I wish we didn't have to talk about this, but there isn't any way around it.

- 找你過來就是要你說說自己的想法。
- The only reason you're here is so I can hear you out.

- 或許會沒有什麼結論，但我們還是先看看能討論到什麼程度。
- This is probably pointless, but let's see how far we can get.

- 有話快說，我沒那個美國時間等你。
- We may as well get going with this—I haven't got all day.

- 你，我就是在叫你！
- I'm calling you out!

粗魯
- 現在就過來跟我說清楚！
- Get in here and talk to me now!

004 如何結束談話

專業
- 和你談話總是這麼開心。
- Talking to you is always a pleasure.

- 真不敢相信我們會聊得這麼愉快。
- I can't believe how wonderfully our talk went.

- 這次的談話對我很有幫助，感謝你。
- This dialogue was very helpful, thank you.

- 這次的討論讓我獲益良多，謝謝你。
- I learned a lot talking to you, thanks for everything.

- 聽君一席話，勝讀十年書。真希望能再多聊一會兒。
- I'm sorry this is over—I was learning a lot.

- 謝謝，今天聊得很開心。
- Thank you, I enjoyed our talk immensely.

- 今天和你談得很開心。
- It was great talking to you.

- 我從今天的談話裡學習到許多。
- I got a lot out of our conversation.

- 真抱歉，我必須先告辭。
- I apologize, but I must leave.

- 我很樂意一直聊下去，但很不巧必須先離開。
- I could listen to you for hours but, alas, I must go now.

- 我很想繼續談下去，但我已經趕不上…
- I'd like to continue with this, but I'm late for...

- 我有急事得先離開。
- I have pressing matters to attend to now.

- 我們就談到這兒吧！
- Let's stop here.

- 抱歉，我還有點兒事情。
- Sorry, but I'm needed elsewhere.

- 很抱歉，我不能再繼續聊下去了。
- I regret I can't pursue this conversation any longer.

- 抱歉，我想就聊到這裡。
- Sorry, I prefer not to continue.

- 我想我們的討論也差不多告一個段落了。
- This is as good a place to end our discussion as any.

- 這麼大的主題很難在一天之內做出結論。
- We won't be able to get anything out of such a big subject in just a day.

- 我能說的都說完了。
- I can't be of any use to you from this point on.

- 我們談點其他事情好嗎？
- Can we talk about something else, please?

- 這件事已經沒有繼續討論的必要了。
- There's no need to discuss this any further.

- 我們已經講得差不多了。
- We've spoken enough about this.

- 就此打住吧！
- Let's stop this right here.

- 再不踩剎車我們就要越扯越遠了。
- Let's pull back before we go too far.

- 這個話題不應該繼續討論下去。
- This discussion isn't appropriate any longer.

- 這個話題已經令我反感。
- This topic disturbs me quite a bit.

- 我們再討論個十年也不會有結果的。
  - We could talk about it for 10 years and we still wouldn't have an answer.

- 再講下去也只是浪費時間。
  - More talk would simply be a waste of our time.

- 我就是不想再講下去了。
  - I just don't want to speak about this now.

- 我對這個話題已經提不起任何勁了。
  - I'm not interested in this topic any more.

- 對我來說，討論已經結束。
  - As far as I'm concerned, this conversation is over.

- 不好意思，要浪費時間其實還有其他方法。（諷刺語氣）
  - My apologies if I choose to not waste my time with this conversation. [sarcasm]

- 和你講話我會跟著腦殘。
  - Talking to you just makes my brain hurt.

- 聽不到…聽不到…（哼歌貌）
  - I'm not listening... [musically]

- 我對你已經無話可說。
  - I have nothing more to say to you.

- 談話結束！
  - This conversation is over!

不專業

- 你最好在講錯話之前閉上你的嘴巴。
  - You'd better shut your mouth before you put your foot in it.

005 如何傳遞消息

好消息

- 這真是天大的好消息，一定會讓你很開心！
- This news is so good, it will blow your mind!

- 這麼好的消息，我實在藏不住！
- This is such great news, there's no way to hide it!

- 我等不及要告訴你一個好消息！
- I can't wait to tell you this amazing news!

- 這正是你朝思暮想的結果！
- This is what you've been waiting to hear!

- 剛剛奇蹟發生了，請你聽我說。
- The best thing just happened—let me tell you about it!

- 你一定不相信我現在要說的這件事。
- You won't believe the news I have for you!

- 你先聽我講完。
- Wait until you hear this!

- 我很高興地向你宣布…
- I'm really pleased to tell you that...

- 這個好消息很難講得清楚，但我盡量。
- I can't do this news justice, but I'll try.

- 這個消息保證讓你開心一整天。
- This piece of information is going to make your day.

- 我不知道該怎麼說，但是…
- I'm not sure how to say this, but...

- 有件事我想你應該要知道。
- I think this is something you should know.

- 我必須委婉地告訴你這個壞消息。
- I'm going to have to let you down easy.

- 恐怕我不得不告訴你關於…的壞消息。
- I'm afraid I have to give you the lowdown about...

- 我想我就和你直說了。
- I'm going to have to break it down for you.

- 我會盡量說得婉轉些。
- Let me put this as gently as I can.

- 我也不想破壞你的心情，但是…
- I don't mean to bring you down, but...

- 我真希望不是由我來告訴你這個壞消息，但是…
- I didn't want to be the one to tell you this, but...

- 我只是個傳話人，必須告訴你…
- Don't shoot the messenger, but...

- 這件事怎麼講都不妥，我想我就直說好了。
- There's no good way to cut the deck, so let me just say...

- 接下來這件事情實在很難以啟齒。
- It's not easy to say what I'm about to say.

- 我實在不想當個帶來噩耗的烏鴉，但是…
- I hate to be the bearer of bad tidings, but...

- 你聽了一定會心情很差，但是…
- You aren't going to like what I have to say, but...

- 我想不到更好的方法告訴你，但…
- There's no good way to tell you this, but...

- 你先坐下吧，這個消息應該會對你打擊很大。
- Are you sitting down? I'm afraid I have news you probably don't want to hear.

- 我很遺憾必須告訴你這個壞消息。
- I'm afraid I have some bad news to report.

壞消息

- 事情很糟，我就直說了。
- Yes, it's bad—here's the scoop.

006 如何闡述觀點

堅定

- 我可以非常明確地告訴你…
- · I can say without equivocation that...

- 我非常確定，…
- · With much conviction, I say...

- 剛剛我說的…完全是根據個人經驗。
- · I'm speaking from experience when I say...

- 我毫無疑問地認為…
- · There is no doubt in my mind that...

- 我保證我知道自己在說些什麼。
- · I can assure you I know what I am talking about.

- 我可以毫不遲疑地告訴你。
- · There is no hesitation in what I am about to say.

- 如果你讓我實話實說，…
- · If you'll allow me to be frank.

- 在我看來，…
- · From my perspective...

- 我只想說，…
- · Let me just say...

- 我想要說的是，…
- · What I am trying to say is...

- 雖然只是自己主觀的想法，但我還是要說…
- · It's just my gut feeling, but let me say...

- 請你讓我闡述一下自己的看法…
- · If you'll allow me to interject for a moment...

- 大家都說得很好，但我想要講的是…
- · That's all good, but what I am trying to say is...

- 我的推論是…
- · My guess would be...

- 有些話在我心裡不吐不快。
- · I feel I have to get this off of my chest.

- 我有些話想說。
- · I think I'd like to say something here.

- 如果我想得沒錯，… · If I am not mistaken...

- 我不太確定，但我認為… · I'm not sure, but I think...

- 或許我是錯的，但是… · I could be wrong, but...

- 我不敢打包票，但是… · I may not have assurance, but...

- 我知道自己常常是錯的，但或許我的意見還是有些值得參考的地方。 · I know I'm usually wrong, but take this for what it's worth.

保留 · 我可以說一下自己的拙見嗎？ · May I add my measly two cents?

007 如何詢問他人意見

禮貌 · 我希望能知道你對這件事的看法。 · I'd love to get your take on this.

- 你有任何意見想要補充的嗎？ · Is there anything you'd like to add?

- 任何意見我都洗耳恭聽，請踴躍發言。 · I welcome all opinions, so please speak freely.

- 你要不要說幾句話？ · Would you like to interject?

- 你對於…的想法是？ · What do you think about...?

- 任何意見我都歡迎。 · I am open to any suggestions.

- 你要不要一起討論？ · Would you like contribute to this dialogue?

- 你對這件事情的看法是？ · What's your view of the situation?

- 如果能知道你對…的看法，對我很有幫助。
- It would help me a lot to know your opinion on...

- 既然你支持…，那麼你對…的看法是？
- As a proponent of [...], would you say that [...]?

- 交換彼此的意見很重要，你有什麼想說的嗎？
- Dialogue is key—what do you have to say?

- 別怕說錯話。
- Don't be afraid of being misunderstood.

- 請自由發言。
- Please speak freely.

- 任何意見都是好的。
- Any input is welcome.

- 如果你不同意，就告訴我其他的想法。
- If you don't agree, show me an alternative.

- 你有何高見？
- What's your best answer?

- 我必須先知道你的看法。
- I can't proceed without hearing from you.

- 請說實話，要不然就請閉上你的嘴。
- Speak your mind or forever hold your peace.

- 你有答案了沒有？
- Have you got an answer or not?

- 既然你一定要說，那就快說吧！
- You're going to say it anyway, so get it over with!

- 有屁快放！
- Just spit it out already!

莽撞

008 如何表達自己的意見

技巧	
· 我會盡量表達我的看法。	· Let me put this as delicately as I can.
· 這件事雖然還有其他面向要考慮，但我想先說…	· While I think there may be more sides to this, let me begin by saying...
· 我非常肯定地認為…	· Without making any concessions, I believe...
· 先不管最後會選擇哪一邊，我想說…	· Without choosing sides, let me say that...
· 雖然很難說清楚，但我必須承認…	· Although it's hard to put into words, I must admit that...
· 在結果還沒揭曉前，我想說…	· Even before all the votes are counted, let me say that...
· 在沒收到進一步消息之前，我要先感謝…	· Before the news comes in, I would like to acknowledge that...
· 在結果尚未揭曉之前，我想說…	· While all the ballots aren't in, I'd like to say that...
· 我知道這是一個棘手／敏感／禁忌的話題，但是…	· I understand that this is a tricky/delicate/taboo topic, but...
· 我的立場是…	· I'm on the side that says...
· 我必須老實告訴你…	· I have to be honest with you and say...
· 請仔細聽我說關於…	· Please hear me out when I say...
· 我想對你坦白，可以嗎？	· I would like to be frank, is that okay?
· 我得直接了當告訴你…	· I have to tell you this directly.
· 我不想得罪任何人，但是…	· I don't want to step on anybody's toes, but...

- 聽著！我有重要的事情宣布。
- Listen, please! I have something of importance to say.

- 我毫無冒犯之意，但我認為…
- With all due respect, I feel that...

- 我們別再拐彎抹角了好嗎？
- Let's not beat around the bush, okay?

- 這件事很難說清楚，所以我就直說了。
- There's no good way to say this, so I'm just going to say it.

直接

- 我們就不用再客套了，直接把事情攤開來講清楚，好嗎？
- Let's stop with the niceties and get everything on the table, shall we?

009 如何尋求更多的資訊

客氣

- 我比較遲鈍，可以麻煩你再多說點嗎？
- I'm probably being dense, but could you say more on this?

- 我希望能更了解你對這件事的態度。
- I would like to know your position better.

- 你能夠證實這件事嗎？
- Would you be able to substantiate this?

- 可否多解釋一下你的立場？
- Can you elaborate on your position?

- 能不能請你再說一遍？
- Would you say that one more time, please?

- 你能不能幫忙指點迷津？
- Would you please shed some light on this?

- 不好意思，你能不能再多說一點？
- I'm sorry, could you tell me more?

- 可否請你多發表一些看法？
- May I ask you to expand on that?

- 我了解你的觀點，只是有些部分希望你能幫忙釐清。
- Your points are well taken, but there are a few things I still need clarified.

- 我想知道事情的所有始末，能不能請你詳細說明？
- I would like to have all the information—would you please elaborate?

- 或許你可以回答我的疑問。
- Perhaps you could clear something up for me.

- 我沒聽清楚，你介意再說一次嗎？
- I'm not sure what I heard—do you mind repeating it?

- 你可以解釋再詳細一點嗎？
- Would you explain this a bit more thoroughly?

- 你的意思不是很明確，可以重講一次嗎？
- Your message was a bit garbled—would you restate it?

- 我可能誤會你的意思了，可以再講一遍嗎？
- I may have misunderstood you—would you repeat that again?

- 我不確定自己完全了解你的意思。
- I'm not sure I understood you very well.

- 告訴我你想說什麼。
- Show me what you mean.

- 你可以提供更多資訊嗎？
- Can you provide more information?

- 我完全不懂你在說什麼，可以解釋清楚嗎？
- I don't understand you at all. Would you clarify?

- 你能不能講清楚你的意思？
- Can you explain yourself more clearly?

- 抱歉，你的重點是？
- Sorry, what was your point?

- 你到底想說什麼？
- What are you trying to say?

- 你是不是有什麼事情沒有告訴我？
- Is there something you're not telling me?

- 拜託說重點。
- Please get to the point.

- 如果沒有更多的資訊，我們無法再討論下去。
- I need more than this to continue our discussion.

- 直接把底牌亮給我看吧！
- Lay your cards on the table where I can see them.

- 給我一個答案。
- I demand an answer.

粗魯 · 有話快說！
- Just spill the beans already!

010 如何澄清自己的觀點

圓滑 · 請讓我換句話說。
- Let me word that a bit differently.

- 抱歉，我想換個方式再講一次。
- I'm sorry, let me say it another way.

- 我有幾個想法可能對你有所幫助。
- I've got a few pointers that might help you out.

- 請讓我以另一種方式表達。
- Allow me to rephrase.

- 這個問題很複雜；我們看看要怎麼一起解決。
- This is a complicated issue; let's see if we can figure it out together.

- 有幾個辦法也許可以讓你有更深入的了解。
- I have a few ideas that may help you understand better.

- 現在情況有點混亂，但我相信我們能攜手面對。
- This is a bit confusing, but I'm sure we can figure it out.

- 我們先把事情理出個頭緒。
- Let's try to make sense of this.

- 我可以幫你釐清事情的始末。
- I can help you comprehend that better.

- 我再用比較淺顯易懂的方式講一次。
- I'll rephrase the information so it is more easily understood.

- 我來解釋一下。
- Let me help you understand.

- 如果有什麼不清楚的地方，請現在提出來。
- If there are any misunderstandings, let's take care of them now.

- 有什麼不清楚的地方，我可以再說明一次。
- If anything is unclear, I'd like to deal with it.

- 讓我先把話講清楚。
- Let me be clear.

- 我們先把事情澄清一下。
- Let's go over it again, for clarity's sake.

- 我們的溝通似乎有點問題，不如先把事情講清楚。
- I sense that we have a problem communicating—let's get this straightened out.

- 如果你不懂我可以再說一遍。
- I can repeat myself if you don't get it.

- 如果你還不明白，我再說得更清楚些。
- If you don't understand, let me clarify.

- 我試著再解釋一次。
- Let me try to make sense of it for you.

- 如果你真的不懂，我可以再解釋一遍。
- If you're having trouble understanding, I can go over it again.

- 如果你真的不明白，我可以用更簡單的方法解釋。
- I can make that clearer if you really need help understanding.

- 如果你不懂，我想你可能要多做點功課了。
- If you don't understand, I guess you'll need to do some research.

- 不懂就問，別亂猜。
- Don't guess when I can easily set you straight.

- 看你一臉茫然，我還是再說一遍好了。
- I guess I'd better clarify what you can't seem to grasp.

- 我現在就再解釋一次你不懂的部分。
- Let me make clear what you clearly don't comprehend.

- 看來這已經超過你能理解的範圍，讓我告訴你吧！
- You're clearly out of your depth, so let me enlighten you.

粗魯
- 你真是無可救藥。
- You're hopeless!

011 如何向他人傾吐祕密

信任

- 我覺得所有事情都能夠和你分享。
- I feel like I could tell you anything.

- 老實說，你是我唯一信得過的人。
- I have to be honest—you are the only one I trust.

- 你是我唯一信任的人。
- You are my sole confidant.

- 我能說知己話的人就只有你。
- You're the only one I can really talk to.

- 我知道在你面前我可以毫無保留。
- I know I don't have to hold back when we speak.

- 你是我第一個想到的人。
- You're my go-to guy/girl.

- 我很重視你的意見。
- I put a lot of stock in your opinion.

- 我幾乎不會告訴別人我的心裡話。
- There are very few people I confide in.

- 我完全信任你。
- I trust you implicitly.

- 我知道你會替我保守祕密。
- I know I can rely on you to be discreet.

- 這件事我可以很放心地告訴你。
- I feel comfortable discussing this with you.

- 如果不相信你，就不會告訴你了。
- I wouldn't say this if I didn't have faith in you.

- 這件事只要你我知道就好。
- This is just between me and you.

- 我不會輕易相信任何一個人。
- Trust is very important to me.

- 認識這麼久，我知道可以相信你。
- After knowing each other for so long, I know I can trust you.

- 老實告訴你…
- To tell you the truth...

41

- 我說的每一句都是實話。
- My word is my bond.

- 你能保守祕密嗎？
- Can you keep a secret?

- 我應該可以告訴你所有事情吧？
- I can tell you anything, right?

- 我現在說的事，不能再傳出去。
- What I tell you here, stays here.

- 我必須告訴你一個祕密，但我不確定這麼做妥不妥當。
- I have to disclose something to you, but I'm not sure if I should.

- 我要你確定我們今天的談話內容絕對不會外流。
- I just want to confirm that these conversations are completely confidential.

- 有些事情你自己知道就夠了。
- There are some things you must keep to yourself.

- 我現在說的話你就當作沒聽過。
- What I'm about to tell you dies here.

- 如果你敢講出去，我們就完了。
- If you tell this to anyone, our relationship is over.

不信任

012 如何轉換話題

客氣

- 除了原本安排的議程之外，我希望能撥點時間討論這件事。
  - With the greatest respect for the agenda, I would like to also discuss this.

- 由於近來發生的一些事情，我想增加幾項討論重點。
  - Because of recent events/updates, I would like to prioritize a few additional points.

- 我希望在做出決定之前先討論另一個議題。
  - Without conceding any points, I would like to address this side issue for a moment.

- 我們應該著眼未來，先把精力放在其他議題上。
  - We should look to the future now and discuss other ideas.

- 今天時間有限，我們先往下討論。
  - Let's move on to something else—we have a lot to cover.

- 我們先進入下一個議程。
  - Let's move on to the next point on the agenda.

- 我們不應該在這個問題上耽擱太久。
  - Let's not dwell on this too long.

- 你介意我們先換個議題討論嗎？
  - Do you mind if we change the subject?

- 為了節省大家寶貴的時間，我們先討論下一個議題。
  - For the sake of everybody's time, let's move on.

- 可以討論下一件事了嗎？時間寶貴。
  - Can we move on? Everyone's time is valuable.

- 我想大家已經講得差不多了。下一個議題是？
  - I think we've thoroughly exhausted the topic—what's next?

- 對了，你周末有什麼計畫嗎？
  - By the way, do you have any big plans for the weekend?

- 你最近有看什麼好片嗎？我打算看場電影。
- I'd like to go to the movies—have you seen any good ones recently?

- 天氣真好，你說對嗎？
- Nice weather we're having, isn't it?

- 你想不想換個話題？
- Wouldn't you rather be talking about something else?

- 可以拜託換個話題嗎？
- Can we please talk about something else?

- 我們別花太多時間在支微末節上。
- Let's not dwell on this unnecessarily.

- 我覺得再討論下去的意義不大。
- I don't think it's constructive to continue discussing this.

- 老實說，我想討論點別的。
- Actually, I'd rather talk about anything else but this.

- 拜託別再講同一件事了。
- Please let's not persist with that.

- 你為什麼要花這麼多時間在這個議題上呢？
- Why do you feel it's necessary to drone on about this?

- 別再歹戲拖棚了。
- Let's not beat a dead horse, okay?

- 可以先把這件事擱一旁嗎？
- Can't we put this topic aside?

- 這件事一時之間也解決不了，別再執著了。
- This isn't a subject that can be solved right now, so let's not even try.

- 現在的討論根本漫無邊際，換下個議題吧。
- Talking about this is like being up a river without a paddle—let's move on.

- 我們根本在兜圈子。
- We're talking in circles.

- 往下討論好嗎？拜託！
- Let's move on, please?

- 拜託，別再講這件事了。
- Please, let's stop talking about this.

粗魯

- 如果你再繼續唸下去，我真的會抓狂。
- If you keep hammering this incessantly, I just might scream.

013 如何表達懷疑

含蓄

· 在我看來…	· It seems to me that...
· 雖然不是很肯定，但我認為…	· I am not positive, but I think that...
· 在我印象中，…	· I am under the impression that...
· 也許不是每個人都知道，但是…	· I don't think it's common knowledge, but...
· 我的感覺是…	· I have the feeling that...
· 我對這件事有其他的想法。	· I'm having second thoughts about this.
· 這整件事有點說不通。	· Something doesn't add up here.
· 我覺得事情有點不對勁，但說不上來是哪裡有問題。	· It seems that something is missing, but I can't quite put my finger on it.
· 我對這件事有些懷疑。	· I am a bit skeptical about that.
· 我不確定我了解你的意思。	· I'm not sure I understand what you mean.
· 我對這件事並不是百分之百肯定。	· I am not 100-percent positive about that.
· 這件事有太多讓我不放心的地方。	· There's a lot about this that I'm not sure of.
· 這整件事根本說不通。	· The numbers just don't add up here.
· 你的結論究竟是怎麼來的？	· Where did you come up with that conclusion?
· 我對你的資料來源感到懷疑。	· I'm not confident in your sources.
· 我不是不相信你，但我的確有點疑慮。	· I'm not trying to insult you; I just have my doubts.

- 我不想多做討論，因為我不相信這是真的。
- I won't belabor the point because I don't think it's true.

- 對於剛剛提到的事情我還是有些疑慮。
- I have my doubts about what was just said.

- 你是真的知道還是自以為知道？
- Do you really know or just think you know?

- 我不知道你的目的到底是什麼。
- I don't know what you're aiming at here.

- 這根本不可靠。
- There's no certainty to that.

- 我高度懷疑這件事。
- I'm unsure about a lot of this

- 我對這事有不好的預感。
- I have a sinking feeling about this.

- 這件事另有蹊翹。
- Something is rotten in the State of Denmark.

- 你是哪根筋不對勁？
- Where did you get that idea?

- 小學生都比你還清楚。
- I've seen clearer thinking from a kindergartner.

- 好吧！隨便你怎麼說 。（諷刺語氣）
- Okay, whatever you say. [sarcasm]

- 看過你的「報告／企劃／作業」以後，我真懷疑你是怎麼平安活到現在。（諷刺語氣）
- After reviewing your [work/plan/operation/idea], it's a wonder you can even tie your shoes. [sarcasm]

直接

如何和失聯已久的舊識重拾連絡

客氣

- 我很懷念我們的友誼，真高興我們又連絡上了。
- I've missed our relationship; I'm so glad we worked things out!

- 能再次和你共事／碰面真是太好了。
- It's so nice to be working together/ hanging out with you again.

- 能重新找回我們過去的交情真是令人開心。
- It's good to restore our relationship and move forward.

- 我很高興我們又可以像以前一樣。
- I am glad this relationship is back on track.

- 我知道你很忙，可能沒空回我上次寫給你的信。
- I know you're busy, so I understand if you missed my last e-mail.

- 我很開心又和你碰上了，我很懷念我們之間的談話。
- I'm so glad we're back together. I've missed our conversations so much.

- 我們兩個都這麼忙，能夠連絡上真是太好了。
- The both of us have been so busy; it's nice to be in touch again!

- 既然我們又連絡上了，不如聊一聊彼此的近況吧！
- Let's take a few minutes to catch up now that I have you back on the line.

- 很遺憾我們失聯了這麼久，其實我一直想找你。
- I'm sorry we grew apart—it was never my intention.

- 雖然之前發生了這麼多的事情，但我們可以重新開始我們的友誼嗎？
- There's a lot of water under the bridge—can we agree to start over?

- 我很期待能夠化解以前的誤會，繼續我們的友誼。
- I'm happy to resolve any issues we've had in the past and look forward to the future.

- 現在我有了你的電話號碼／電子信箱，就能隨時保持連絡了。
- Let's reconnect now that I have your phone number/ e-mail address again.

- 我們找時間碰個面，聊一聊各自的近況吧！
- Let's get together and discuss what we've missed out on.

- 真抱歉，我之前真的忙到分身乏術。
- I'm sorry, I guess I just got too busy/overwhelmed to deal with everything.

- 我希望有空能和你聊聊最近的生活。
- I hope you can take a few minutes to catch me up on what you've been doing.

- 我覺得我們之前似乎對彼此有誤解。
- I feel that you and I somehow got lost in translation.

- 我們應該握手言和，把之前的誤會講清楚。
- We should reconcile and talk things over.

- 過去的事就算了，我們重新開始吧！
- Let's let bygones be bygones and start over.

- 我們應該面對面把事情講清楚。
- We need to get together again and talk things out.

- 你和我竟然曾經失聯，真不知道是怎麼發生的。
- You and I clearly fell through the cracks—it's hard to understand why.

- 我們根本不應該失去聯絡。
- We should have never lost touch.

- 如果你願意不計前嫌，我也願意。
- I'm willing to let it go if you are.

- 你來決定我們的關係吧！
- The ball is in your court now.

粗魯

- 過去的事就讓它過去吧！
- Let's just move on already!

015 當你回答完一個問題之後

風度

・我有回答到你的問題嗎？	・Did that answer your question?
・這是你需要的答案嗎？	・Was that what you needed from me?
・這是你在尋找的答案嗎？	・Was that the kind of answer you were looking for?
・這個回答你滿意嗎？	・Was that answer satisfying?
・我很高興自己能回答到你的問題。	・I'm glad I was able to provide a good answer.
・這個答案有道理嗎？	・Did that make any sense?
・你清楚我的邏輯嗎？	・Are you following my train of thought?
・剛剛的答覆有些冗長，謝謝你聽完。	・Thanks for listening; that definitely wasn't a short answer!
・剛才的答覆很長，謝謝你的耐心。	・That was a long answer; thanks for your patience.
・如果我講得不清楚，請直說。	・If I wasn't clear enough, please let me know.
・你聽得懂我的回答嗎？	・Were you able to grasp what I was saying?
・如果你聽不太懂，我很樂意再說一次。	・If you didn't understand everything, I would be glad to go over it again.
・如果你不是很了解，我可以再解釋一次。	・If you didn't catch all of that, I could go over it again.
・嗯，有人似乎沒有專心在聽喔！	・Hmm, someone wasn't listening!
・你聽不懂哪個部分？	・What part of the answer didn't you get?

- 我的回答清楚嗎？
- Am I being clear or not?

- 如果你聽不懂，我也愛莫能助。
- I can't help it if you didn't understand me.

- 我沒空詳細解釋，時間寶貴！
- I don't have time for more details; time is money.

- 這是我的答案，信不信由你。
- That was my answer—take it or leave it.

消極

- 該說的我都說了，你要怎麼想隨便你。
- That's all I'm saying—I really could care less what you think/how you interpret it.

016 當你不想回答問題時

圓滑

- 這個問題真的很難回答。
- There's probably no easy answer to that.

- 這個問題可以有好幾種解釋。
- There is more than one way to look at that.

- 你問的問題並沒有一個直接的答案。
- There's no straightforward answer to what you're asking.

- 讓我先想想再回答你。
- Let me think on that and get back to you.

- 我真的不知道該怎麼說。
- I just don't know what to say.

- 抱歉，我真的無話可說。
- Sorry, words escape me at the moment.

- 我對這件事的了解還不足以回答這個問題。
- I don't know enough to give a definitive answer either way.

- 這個問題並沒有一個簡單的答案。
- There isn't a simple answer to that question.

- 這件事沒有辦法用三言兩語解釋清楚。
- There are no simple explanations.

- 關於這件事的說法眾說紛紜，我持保留態度。
- There's more than one school of thought, so I can't take a hard stand.

- 這個問題超出了我的能力範圍，需要專家來解答。
- This is a specialized topic and I'm certainly no expert.

- 我覺得這件事情還有討論的空間。
- I think that it's arguable.

- 我目前沒有足夠的資訊，無法提供確切的答案。
- I don't have enough information to give you an intelligent/informed answer.

- 我對事情的始末並不了解。
- I don't know all the details about that.

- 這件事說來話長。
- That could take hours to explain.

- 我無法回答我不確定的事。
- I'm not sure, so perhaps it's best not to answer at this time.

- 這個問題牽涉到非常專業的知識。
- You would need specialized knowledge to understand.

- 這個問題很複雜，老實說，回答了你也不一定能理解。
- The answer is complicated, and honestly, you probably wouldn't understand anyway.

- 這牽涉到技術層面，我不確定你有足夠的專業知識來理解。
- It's quite technical; I'm not sure you have the knowledge to understand.

- 雖然我很想多花點時間在這個問題上，但真的沒辦法。
- As much as I'd love to spend time on this issue, I can't.

- 這個問題事關重大，我必須審慎思考後才能回答。
- The question is so important, I want to take some time before answering.

- 我現在無法回答這個問題。
- I'd rather not answer that right now.

- 我不想碰觸這個話題。
- I don't wish to enter into this conversation.

- 我沒有資格回答這個問題。
- I am not proficient enough to answer.

- 這已經超出我的專業領域。
- This is outside my area of expertise.

- 我沒有辦法回答這個問題，換下個議題吧！
- I have a tough time talking about this— let's move on to something else.

- 我對這件事沒什麼意見。
- I have no opinion on the matter.

- 我的原則是不在朋友／家人／同事前談論這件事。
- I make a rule never to speak about this with my friends/family/coworkers.

- 我不想談這件事。
- I prefer to not talk about it.

- 我不打算參與這個討論。
- I prefer to stay out of this conversation.

- 就算我有任何意見，也不會在這裡透露。
- If I had something to say about this, I certainly wouldn't say it here.

- 這個時候你的問題一點也不重要。
- Your question isn't important in times like these.

- 我不是要污辱你的智商，但這事沒你想的那麼簡單。
- I don't want to insult your intelligence, but this isn't as simple as you think.

- 如果你尊重我，就不會問這種問題。
- When you respect someone, you don't ask such questions.

- 我不置可否。
- I can neither confirm nor deny that.

- 無可奉告。
- I decline to comment.

- 我有保留沈默的權利。
- I plead the Fifth.

- 你去問別人吧！
- You'll need to ask someone else.

直接
- 我不知道。
- Beats me!

當別人請你重講一次

隨和

· 沒問題，我很樂意。	· No problem; I'd be happy to!
· 好啊！	· Gladly!
· 很抱歉我剛剛講得太快／太小聲，我很樂意再重複一次。	· I'm sorry if I went too fast/spoke too softly—I'll gladly say it again.
· 我知道，這裡的音響效果很差（玩笑語氣）。	· Yes, the acoustics here are terrible. [joking]
· 我再重申一次…	· I'd like to reiterate that...
· 我再說一遍。	· I'll say it once more.
· 如果你需要，多講幾次也沒關係。	· I'll say it a thousand times if you need me to.
· 只要能解釋清楚，我不介意多講幾次。	· I don't mind repeating myself, as long as it makes things clearer.
· 只要你有需要，我願意再說一次。	· I'm willing to elaborate if that's what you need.
· 你顯然沒聽清楚，我剛剛說的是…	· You obviously didn't hear me correctly—here's what I said.
· 我需要用麥克風嗎？	· Does anyone have a microphone?
· 我再認真地重講一次。	· Once more, with feeling.
· 這次就算了，我通常不會再說一次。	· It isn't my habit to repeat myself, but obviously it's necessary.
· 你可以等會議記錄出來後自己看。	· You can wait for the transcript if you want.

- 雖然是浪費時間，我還是再說一次。
- I'll say it again, even though we're wasting everyone's time.

- 抱歉，但我沒時間再重複一次。
- I apologize; I don't have time to go over this again.

- 或許我們可以改天／稍後／之後再找時間討論這件事。
- Perhaps we can go into that again some other time/later/afterward.

- 抱歉，我沒空再講一次。
- Sorry, there's no time to repeat myself.

- 這裡的音響效果很差，再講一次可能也是白費力氣。
- The acoustics here are so bad, I doubt that repeating myself will do any good.

- 我擔心重述我的想法只會讓事情變得更複雜。
- I'm afraid that restating my position will only serve to confuse the issue.

- 我不會再說一次，否則會耽誤到其他事情。
- I won't say it again because it's only going to hold things up.

- 我剛剛都說得很清楚了，沒必要重複。
- Everything I just said was crystal clear and doesn't need repeating.

- 這樣會佔用到其他人的時間吧！
- Doing that would be inconsiderate of the others.

- 如果你剛剛沒有認真聽，我也沒必要重講一次。
- If you had listened in the first place, I wouldn't have to repeat myself.

- 我剛剛講第一遍的時候你就應該專心聽才對。
- You should have been listening more closely when I said it the first time.

- 該講的我都講了。
- I told you enough about what I think.

- 我再說最後一次，所以仔細聽著。
- I'm only going to say this once more, so pay attention this time.

- 我認為完全沒有再講一次的必要。
- Saying it again is completely unnecessary.

- 拜託你把耳朵打開好嗎？
- Please remove the wax from your ears.

・看來我每一次都要重講第二
遍。（諷刺語氣）

・Repetition is my greatest ally, it seems.
[sarcasm]

・你聾了嗎？

・Are you deaf?

辦公室厚黑學

說話不僅是傳遞訊息，深處攸關個人未來發展的職場，除了要避免一般人常犯的錯誤外，掌握關鍵技巧，就能展現出領袖風範。

溝通哪些事

- 如何要求加薪
- 如何要求休假
- 如何拒絕老闆
- 如何避免碰觸私人話題
- 如何提出一個私人或令人尷尬的話題
- 如何提升他人信心
- 如何要求單獨談話
- 當你需要他人的注意力
- 如何分配任務給其他人
- 如何整肅會議秩序
- 如何結束會議
- 如何讓談話重新聚焦
- 如何提議一項行動或方案
- 如何推延一項任務
- 如何推延一場對話
- 如何推延做決定的時間
- 如何團結眾人
- 如何稱讚上司
- 如何鼓舞員工
- 當員工表現退步時
- 如何開除員工
- 如何強調事情的急迫性
- 如何緩和事情的節奏

> **心懷大志的人才能做大事**
> ——佛南多·佛羅倫斯（Fernando Flores，智利政治家）

　　不管是基層員工或是中高階主管，和公司同事間的溝通都十分重要。尤其當你打算努力爬上高位，如何在言談間展露領袖特質就更為關鍵。下面這幾點，即使你只是個剛進公司的小員工，也應該銘記在心：

 正面語言

　　保持正面態度是升遷的一大關鍵。積極的人多半會主動採取行動，也比較受同事歡迎。就算是討論一件帶有負面色彩的事，正向的談話通常能得出對策。相反地，態度消極的人一般只會不斷抱怨，又不提出任何有助於問題的想法或建議。最後只會被同事當成自怨自艾的可憐蟲，和領導特質沾不上任何關係。

2 三思而後言

　　在開口之前千萬要先想好自己要說的話。話脫口而出後才發現傷害到某個人或是洩漏了重要祕密，應該是每個人都犯過的錯誤。因此講話之前請先在腦海裡檢查一遍，可以避免說出讓你事後追悔莫及或是言者無心的話（這是政治人物常犯的毛病）。難怪法國有句俗話：「開口之前先在嘴裡捲七次舌頭。」

❸ 一次一件事

　　現在的工作環境步調飛快，很多人習慣事情做到一半，就匆匆著手下一件事。把腳步放慢，好好處理完手上的事情再進行下件事。如此，別人會認為你很細心謹慎，並對你所做的事情更有參與感。

❹ 演說前做準備

　　一個好的演說者很清楚，演說前的準備其實比演說的內容更重要。事先把演講內容的重點列出來，有邏輯地整理出順序。小名片卡是很好用的工具，簡單整理而且方便更動順序。如果你不習慣對一大群人講話，可以先在朋友和家人面前彩排練習。成功的彩排能夠幫助你檢查內容、用詞遣字、節奏快慢等。每次演講時，請記得先帶出整個演講的大綱以及長度。最後務必用正面積極的語句結束演說。

❺ 除了講，還有寫

　　不管是紙本信件、電子郵件，或者是部落格文章，書寫的時候請遵守下列幾點：

　　◆時時提醒自己讀者是誰。語意清晰，但別過度正式，除非你是寫信給英國女王。只要找出你和讀者之間的共通點就沒錯了。

　　◆注意自己的文法、句型和修辭。比方說，和自己的伴侶互傳訊息時當然可以用縮寫或口語，但對外的溝通，你的文字就是個人形象的一部分，不能錯字連篇。

　　◆格式很重要。條列式重點會讓內容更清楚易懂。另外可適度搭配不同的字型、粗體、斜體或底線等變化。

018 如何要求加薪

專業

- 請問您可以撥出一些時間來討論我的薪資報酬嗎？
- Could you please take a moment and review my current level of compensation?

- 自從加入公司以來，我的薪水一直沒有調整。請問可以討論一下嗎？
- My salary hasn't changed since I began working here. Can we look into that?

- 我的職責增加，卻沒有反映在薪水上。
- I've taken on additional responsibility but don't have the salary to show for it.

- 現在部門人數縮編、工作量加重，我認為應該加薪。
- We're a much smaller team than when I started, so I need a raise to compensate.

- 我做了一些研究，發現自己的薪水遠低於市場行情。
- I did some research and I'm not making anything close to my current market value.

- 上次我們討論薪水已經是一年多前的事了。
- It's been more than a year since I had a salary review.

- 我覺得自己目前的工作表現比以前出色許多，但薪水卻在原地踏步。
- I think my work is the best it has ever been, but I'm still making the same salary.

- 可以討論一下該如何針對目前的物價水準調整薪水嗎？
- Can we at least discuss a cost of living adjustment?

- 我不懂為什麼我的薪水一直沒有調整，你知道原因嗎？
- I can't think of any reason why I should be making the same money, can you?

- 我希望能了解公司幫員工加薪的條件是什麼？
- Let's discuss what needs to happen in order for me to get a raise.

- 加薪雖然不能消除工作壓力，但絕對是一大鼓勵。
- A raise won't get rid of the stress, but it sure will help me feel better.

・我的薪資低於其他同事，希望能找個時間和你談一談這件事。

・Everyone else is making more than I am; I hope we can discuss that.

・我需要加薪。

・My salary needs an upward adjustment.

・我需要做一次績效檢視和調薪，請問您什麼時候有空？

・I'm due a performance review and a raise. When can we make than happen?

・根據我的工作表現，加薪是應該的。

・I don't think a raise would be unwarranted, given all I do here.

・我一個人做兩份工，我要更多的錢。

・I'm doing the work of two people. I simply need more money.

・我覺得公司應該要幫我加薪，如果沒有，我會很失望。

・I think I deserve a raise, and I won't be happy unless you agree.

・如果你滿意我的工作表現，請給我實質的獎勵。

・If you are happy with my work, you need to show that to me in concrete terms.

・如果薪水再不調整，我將提出辭呈。

・I'm going to quit unless I get a salary adjustment.

・不加薪我就不幹了。

・Either give me a raise or I quit.

・我真的做不下去了；要我留下的唯一辦法就是給我更多錢。

・I can't take this job anymore; if I stay I'll need to make more money.

挑釁

・如果你不當場幫我加薪，我馬上走人。

・If I don't get a raise right now, I'm walking out.

019 如何要求休假

溫和

- 如果你不介意，我想和你討論一下休假的事。
  - If it's okay with you I'd like to discuss taking some time off.

- 我打算休假，請問這幾個日期可以嗎？
  - I was thinking of taking a vacation—do these dates work for you?

- 請問現在這個時候方便休假嗎？
  - Would this be a good time to request some time off?

- 請問該如何申請休假？
  - What form do I need to fill out to request vacation?

- 請問你上個月休假是怎麼申請的？
  - What form did you fill out when you went on vacation last month?

- 如果公司有標準的休假申請流程嗎？我想休幾天假。
  - If there is a standard vacation request, I'd like to fill one out.

- 家裡出了些狀況，我需要請一個星期的假。
  - I'm having a family emergency so I'll need to take a week off.

- 抱歉，臨時有重要的事情必須處理，我要請一個禮拜的假。
  - A sudden emergency has come up; I need a week off to tend to it.

- 家裡有人過世了，我必須請喪假。
  - A family member has passed away and I need bereavement leave.

- 我的預產期快到了，我要開始請產假。
  - I'm going to have a baby so I'll need to request some maternity leave.

- 我這陣子健康出了點問題，需要請幾天病假。
  - I've been having some medical problems so I will need to take some of my sick leave.

- 請問任職多久才能開始申請休假？
  - How long do I have to be on the job before I qualify for time off?

- 最近工作壓力太大，我非得休假不可。
- Because of all the stress lately I need to schedule a mental health break.

- 我需要放假。請問要怎麼申請呢？
- I deserve a vacation. Can you tell me how to request one?

- 大部分的同事工作到這個時候都會請個假，我想我也需要。
- Most people who have worked here this long get time off; I think I should, too.

- 我覺得沒有不能放假的理由，這是我應得的。
- I don't see why I can't take some vacation time. I've earned it.

- 我想先讓你知道我打算在這幾個日子請假。
- Just wanted to give you a heads-up: I'm going on vacation on these dates.

- 假是一定要請，我只是在考慮該請一個禮拜還是兩個禮拜？
- The only question I have is, should I take one week off or two?

- 到了我這樣的資歷，請假應該不需要任何人的批准吧？
- I don't believe I'm at the level where I need to ask permission to take some time off.

激動

- 我還有假，不管你說什麼我都會把假休完的。
- I've earned some time off and I'm going to take it, no matter what you say.

020 如何拒絕老闆

婉轉

- 好，但是我現在應該先處理哪一件工作呢？
- Okay, but which of these tasks do you want me to finish first?

- 那你交代我的另外三件事該怎麼處理？
- What about the other three tasks you've given me?

- 我很想馬上處理，但是手上正在忙其他的事。
- I wish I could help with this but I am tapped out at the moment.

- 有可能請其他人先幫忙嗎？我真的分身乏術。
- Is there someone else you could call on? I have so many plates in the air right now!

- 如果接下這個案子，我怕工作品質不會太好。
- If I say yes I'm afraid my work quality will suffer.

- 抱歉，我不覺得自己還有辦法接下更多任務。
- I can't imagine taking on that much responsibility, sorry.

- 我也希望自己有三頭六臂，但我畢竟只是一個人而已，對不起。
- I wish I had five heads but I'm only one person, I'm afraid.

- 以目前的工作量來看，你的要求有些不合理。
- I'm not sure that your request is entirely reasonable, given how hard we are all working.

- 這以前是兩個人的工作，一個人真的做不來。
- Two people used to do that job—can one person really do it alone?

- 如果薪水能相對調整，我很樂意接下更多工作。
- I'd be happy to, as long as my salary is adjusted to reflect the additional workload.

- 我們可以多請個人／助理來嗎？
- Can we look into hiring someone else/ an assistant?

- 你不應該把所有工作都堆到我頭上。
- I don't think it's right for you to pile all this work on me.

- 雖然你有權力叫我做這些事，但不表示這是合理的。
- It may be within your authority to ask me to do that, but I don't think it's a smart move.

- 你要我做的這些事完全不在我的工作責任範圍內。
- That's not in my pay grade/part of my job description/something I am going to do.

- 這件事或許應該由你自己去處理。
- Maybe you should handle that yourself.

- 如果是你，做得完這麼多工作嗎？
- Do you think you could handle that much work?

- 你的要求一點也不合理，請不要再提了。
- You've exceeded your boundaries, and I'd like you to stop.

- 我拒絕被當做機器人（奴隸）一樣使喚。
- I am going to have to object to being treated like a robot/slave.

- 這已經超出了我可以容忍範圍，我拒絕。
- This is completely out of bounds and I'm not going to stand for it.

- 把自己不想做的事推給員工做，公平嗎？
- It's unfair to ask something of your employee that you're not willing to do yourself.

- 我拒絕。
- I'd rather not.

- 我不是你的奴隸。
- I am not your slave.

直接
- 你並不是我的主人。
- You're not the boss of me.

021 如何避免碰觸私人話題

有禮

- 我現在忙得快喘不過氣了，可以晚點再聊嗎？
- I'm up to my neck right now—can we talk later?

- 我們可以下班後另外見面討論嗎？
- Can we meet after work to discuss this?

- 抱歉，但我不太想談這件事。
- I'm sorry, but I'd just rather not talk about that.

- 這對我來說是個不愉快的話題，抱歉。
- That's kind of a sore subject for me, sorry.

- 這件事牽涉到我的個人隱私／家庭，不好意思。
- That's just too upsetting/personal/close to home, sorry.

- 我不習慣在工作時討論這方面的事。
- I'm just not comfortable talking about that at work.

- 等更恰當的時間、地點，我們再來談這件事。
- This is probably not the best time or place to talk about this.

- 或許這件事情我們可以晚點再聊。
- Maybe we should talk about this later.

- 在這裡講其他人會聽到，晚點吧！
- I think other people can overhear us—let's talk later.

- 我現在忙得頭昏腦脹，或許午餐時再聊？
- I'm under the gun today. Maybe we can talk at lunch.

- 你不覺得工作的時候講這些事情，似乎不太恰當嗎？
- I feel this is a bit too personal to talk about at work, don't you?

- 這件事太過私人了。
- That's simply too personal to address here.

- 我不想談這件事。
- I don't feel like talking about that.

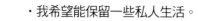

- 我希望能保留一些私人生活。
- I prefer to keep my private life private.

- 我不喜歡在工作場合討論這些事。
- That's not something I like to discuss at work.

- 我看起來像是有空聊這些事嗎？
- Do I look like I have the time to talk about this?

- 我喜歡把工作和私人生活區分清楚。
- I prefer to keep my work life and private life separate.

- 這件事沒什麼好聊的。
- There's no reason to bring that up.

- 我有說過我想聊這件事嗎？
- When did I ever give you the impression that it was okay to talk about this?

- 哇，這個話題太不恰當了吧！
- Wow, this is so not appropriate.

直接

- 我如果想聊這些事，怎麼可能不找你？（諷刺語氣）
- Sure, you're the person I would want to discuss this with. [sarcasm]

022 如何提出一個私人或令人尷尬的話題

有禮

- 如果我告訴你一件事，可以替我保守祕密嗎？
- May I tell you something in confidence?

- 我們可以聊一下嗎？我實在不知道還可以找誰講這件事。
- I don't know who else to turn to—can we talk for a moment?

- 我可以把門關上嗎？我想聊一些私人的事。
- May I close the door? I have something personal I need to share.

- 我想和你講一些很私人的事，你現在有空嗎？
- I've got something to share that's fairly private. Do you have a moment?

- 我真的很想告訴你一些事。
- I really need to talk to you about something.

- 這麼說並不是懷疑你，但我真的希望你能保守祕密。
- This is most unpleasant, but I really appreciate your tact and discretion.

- 這件事關係到個人隱私，但我相信你知道怎麼處理。
- Admittedly this is kind of personal, but I think you can handle it.

- 這個話題有些尷尬，但我知道我能夠信任你。
- I know this is embarrassing, but I know I can trust you.

- 我必須和你談一談最近發生的一件事。
- Something's come up that I need to talk to you about.

- 呃，我必須和你坦白一件事。
- Ugh, I need to confide in you for a moment.

- 這件事雖然很私人，但你是我的朋友。
- This may be a little personal, but I consider you a friend.

- 你現在有幾分鐘的空檔嗎？我真的需要和人聊一聊。
- I've never needed to talk this badly. Do you have a minute?

- 你是唯一能了解的人。
- You're the only person here who will understand.

- 我告訴你的事請不要說出去。
- Please keep this under your hat when I tell you.

- 我現在要和你講一件事，但你必須保守祕密。
- I'm going to tell you something but you need to keep it a secret.

- 嚴格說來，這件事並不適合在任何時候說，所以我就直說了。
- There's never a good time or place to discuss this, so I'm just going to say it.

- 不管我說了什麼，聽過就忘了。懂嗎？
- What I tell you here dies here, okay?

- 這真的很尷尬，你確定要聽？
- This is pretty embarrassing—you sure you wanna hear about it?

- 我知道這麼說不太得體，但是…
- I know this is inappropriate, but...

- 我希望這麼說不算是大嘴巴，但是…
- I hope this isn't TMI, but...

直接

- 拜託，你先聽我講完這件事再說吧。
- Yuck, wait until you hear this.

023 如何提升他人信心

誇張

- 你對這件事真的很拿手，我該要擔心自己的飯碗了。
- You're so good at what you do, I should be fearful of losing my job!

- 我要怎麼做才能趕在你之前把事情處理好呢？
- How did I ever get anything done before you arrived?

- 有你不會的事嗎？
- Is there nothing you can't do?

- 我從來不擔心你負責的部分。
- I know I never have to worry about your work.

- 我完全相信你的能力。
- I have complete confidence in your abilities.

- 你生來就是吃這行飯的。
- You were born to do this.

- 你是這個工作的不二人選。
- You're the perfect person for this job.

- 我知道你做得到。
- I know you can do it.

- 你是這裡最好的員工之一。
- You are among the best workers here.

- 你處理事情的能力真是令人欣賞。
- Your ability to get things done is admirable.

- 每個人都覺得你很厲害。
- Everyone knows how capable you are.

- 你真的很有天分耶！
- You have a real talent for this.

- 這可能是你職涯的最佳代表作吧！
- You can make this your best work ever.

- 大家都知道你能力很好。
- Everyone knows how good you are.

- 你不會搞砸的。
- You can do no wrong here.

- 你從來沒讓我失望過。
- You haven't disappointed me yet.

・你什麼時候失手過？	・When have you ever dropped the ball?
・你是我最好的人選。	・You're my best choice at the moment.
・我知道只要你有心，絕對做得到。	・I know you can move mountains if you put your mind to it.
・我知道你的能力，做這件事綽綽有餘。	・I know you're equal to the task.
・這正是你上場表現的時候。	・This is the time to really make it count.
・這是你展現實力的機會。	・Here's your chance to show us what you're made of.
・你以前完成過比這還艱難的工作。	・You've done more challenging tasks in the past.
・別擔心，你知道你可以的。	・Don't worry about it—you know you're good.
・大家都告訴我你能力很好，別讓他們失望了。	・People have told me you're good, so don't let them down
・最重要的是要相信你自己。	・What's important is that you believe in yourself.

冷靜

日常對話

辦公室厚黑學 2

處理衝突與憤怒

展現外交手腕

緊湊時刻

解決問題

表現風度禮貌

掌控棘手職輯的帝王學

024 如何要求單獨談話

禮貌

- 如果方便的話，我想和你私下談談。
- I'd rather discuss this in private if that's okay with you.

- 請問我能夠和你私下聊一聊嗎？
- May I talk to you in private, please?

- 我想這件事還是私下單獨談比較妥當。
- I think it would be best to discuss this in private.

- 我不希望任何人聽到我們的談話。
- I would rather not risk anyone overhearing this.

- 我需要和你單獨聊一下。
- I need to see you privately for a minute.

- 我們可以找個安靜的地方談談嗎？
- Can we go somewhere private to discuss this?

- 我一定要和你單獨談談。
- I simply must talk to you alone.

- 我要講的完全是個人私事。
- I'm afraid this is a private matter.

- 我想和你討論個機密。
- I have a confidential matter to discuss.

- 我們找個比較安靜的地方吧！
- Let's go somewhere quieter/more private.

- 你能放下手邊的事和我出來一下嗎？
- Can you take a break and come with me?

- 我們可以找個隱密的地方聊一聊嗎？
- Can we find a quiet spot for a discussion?

- 我想我們應該在比較私密的場合討論這種事。
- I believe this kind of discussion requires a bit more privacy.

- 我想和你單獨談話，我們另外找個地方吧！
- This is a two-person conference; let's go elsewhere.

日常對話

② 辦公室厚黑學

處理衝突與僵局

國際外交手腕

控制時間

解決問題

表現處事圓融

掌控確實訊息的管道學

- 請跟我到我的辦公室。
- Please follow me to my office.

- 請跟我來。
- Walk with me, please.

- 這是你和我的事，其他人不用知道。
- This isn't an open forum; this is between you and me.

- 我不會在任何人面前提這件事的。
- I won't discuss this in front of anyone else.

直接

- 除非只有你和我，否則我一個字也不會提。
- I won't speak of the matter any further unless we're alone.

025 當你需要他人的注意力

客氣

- 不好意思打擾你，但可以麻煩你先注意我這邊嗎？
- I'm sorry to interrupt, but may I have your attention please?

- 我真的不想打擾你，不過還是要請問一下你的意見？
- I hate to be a bother—what do you think of this?

- 我可以佔用你幾分鐘問一些問題嗎？
- May I bother you for a few minutes with my inquiry?

- 我有個簡短的問題要請教你。
- I have a very short question for you.

- 我一定要和你討論幾件重要的事。
- I must go over some crucial points with you.

- 你有在專心聽嗎？
- Do I have your full attention?

- 你聽見我的話了嗎？
- Can you hear me okay?

- 我不確定你有沒有在聽。
- I'm not sure if I have your full attention.

- 這件事很重要，請專心聽。
- This is important—please pay attention.

- 我正在解釋很重要的事情。
- I'm explaining important things to you.

- 如果你專心，就能更明白我在說什麼。
- I think you'll hear me better if you pay attention.

- 你看起來沒在聽我說話。
- You don't seem very attentive to what I'm saying

- 很明顯地，你沒在聽我講話。
- You're clearly not focusing on what I'm saying.

- 我正在對你解釋，但你完全沒在聽。
- I'm explaining things to you and you don't seem to be listening.

- 不好意思，你好像有點分心。
- I'm sorry, but you seem a little distracted.

- 我覺得你沒在聽我說話。
- I feel like I'm not being heard.

- 你有聽進去我的話嗎？
- Am I not getting through to you?

- 我不確定你有專心聽我說話。
- I'm not sure you're listening attentively to me

- 請注意聽，我要說的事情很重要。
- Listen up—what I have to say is very important.

- 你可以給我一分鐘嗎？
- Will you please pay attention for a minute?

- 你真的對我說的感興趣嗎？
- Are you at all interested in what I'm saying?

- 是不是每次都要我提醒，你才會專心聽我講話？
- Do I always have to ask you to listen to me?

- 你知道我正在和你說話嗎？
- Are you even aware that I'm talking to you?

- 我覺得我根本是在自言自語。
- I feel like I'm talking to myself, here.

- 我看得出來你興趣缺缺。
- I have the feeling that you're not very interested.

日常對話

② 辦公室厚黑學

處理衝突與情緒

展現外交手腕

諷刺時刻

解決問題

表現風度禮貌

掌控發言權的帝王學

・你到底想不想聽我說這件事？	・Are you even a tiny bit interested in what I'm saying?
・我根本是在對空氣說話吧？	・Am I speaking to the wall, here?
・我和你說話的時候請看著我。	・Please look at me when I'm talking to you.
・等你能給我五分鐘的時候再來找我。	・Let me know when you can give me five minutes of your time.
・和你說話根本是浪費時間。	・Talking to you is a waste of my time.
・對牛彈琴都比跟你說話好。	・I could communicate better with a rock.
・你有多久沒有清耳朵了啊？（諷刺語氣）	・Cleaning your ears might help. [sarcasm]
・請問，有人在家嗎？（諷刺語氣）	・Hello, is anybody home? [sarcasm]
直接　・你是過動症發作了嗎？	・Did someone forget their ADD meds?

026 如何分配任務給其他人

禮貌　・你能不能幫幫忙去…	・Would you be so kind as to...
・你想做這件事嗎？還是希望我來？	・Do you want to do it, or would you like me to?
・你想參與這件事嗎？	・Would you like to step in, here?
・我這邊有個計畫你可能會感興趣。	・I have an opportunity you may be interested in.

- 你可以負責處理這件事嗎？
- Is there any way you could take care of this?

- 你還願意接下其他工作嗎？
- Are you open to the idea of additional work?

- 可以請你幫個忙嗎？
- Will you please help me out?

- 可否麻煩你花幾分鐘去…？
- Would you take a few seconds to...?

- 你還有辦法接另一個新任務嗎？
- Are you available to take on something new?

- 你可以幫我處理這件事嗎？
- Would you please handle this for me?

- 可以請你幫我個忙嗎？
- Would you do me a favor, please?

- 有件事想請你處理，應該花不了太多時間。
- I have something to ask of you that shouldn't take too long.

- 你能不能考慮接下這個新任務？
- Would you look into this new assignment, please?

- 你介不介意…？
- Would you mind...?

- 這件事花不了幾分鐘。
- This won't take but a moment.

- 如果你能處理一下這件事，我會很感激。
- It would mean a lot to me if you took this on.

- 我手頭上有許多任務要分配出去。
- I have a heavy responsibility to delegate.

- 從現在起你多了一項新工作。
- Here is one more task to add to your list.

- 要不要接手這項任務？
- Care to take a crack at it?

- 如果你能…，我會加發績效獎金。
- I'll pay you a bonus if...

- 如果你能處理好這件事，我就幫你加薪／升職。
- If you can handle this, there's a little extra money/a promotion in it for you.

- 這件事需要在〔什麼時間〕前處理妥當。
- The following needs to be done by [date].

- 我能夠把…交付給你嗎？
- Can I trust you to...?

- 我們需要在〔什麼時間〕前完成。
- We need to have this done by [date].

- 希望你能在〔什麼時間〕前完成這項任務。
- I expect you to have this finished by...

- 是不是每次都要我交代，你才會去做這些該做的事？
- Do I always have to ask you to do things that need to be done?

- 我建議你多分擔一點工作。
- I suggest you take on more responsibility here.

- 你還沒完成那項工作嗎？
- Do you mean that you haven't done it already?

- 這件事不是昨天就應該要完成了嗎？
- Wasn't this due yesterday?

- 我還在等那份…。（諷刺語氣）
- I'm waiting... [sarcasm]

- 我不是在問你要不要做，是叫你去做。
- I'm not asking you to do this, I'm telling you.

- 如果你不想做，那只好請你另謀高就。
- I guess there's always the option of feeling the door hitting you on the way out.

直接

77

027 如何整肅會議秩序

嚴肅

· 可以請每個人回到自己的座位上嗎？	· Would everyone be so kind as to take their seats?
· 我請大家過來這裡是要討論很重要的事。	· I've gathered everyone here to talk about something very critical.
· 請大家就座。	· Please, everyone take your seats.
· 請問，我們可以開始正式討論了嗎？	· Let's get down to business, shall we?
· 那麼我們開始切入今天的主題吧！	· Let's go over the purpose for our meeting today.
· 我們開始今天的討論吧！	· Let's open this up for debate.
· 這是今天會議的流程。	· Here's our order of business for today.
· 今天會議的流程是…	· Our agenda for today's meeting is...
· 請大家保持會議秩序！	· Let me call this meeting to order.
· 今天我們之所以召開這個會議是因為…	· The reason we're here today is...
· 首先，我們先從今天會議的流程開始。	· Let's begin this by first outlining our order of business.
· 等各位準備好紙筆之後，我們的會議就正式開始。	· If you have your notepads ready, let's get started.
· 已經到了會議開始的時間，請大家注意我這裡。	· It's time to start, so please give me your undivided attention.
· 這次討論的主題是…。	· The topic(s) for this session is...
· 讓我們一起好好深入探討今天的主題。	· Let's dig right into the subject of today's meeting.

- 我希望我們能先從複雜的議題開始討論。
- I'd be happy if we could tackle the complex issues first.

- 我們先從今天預定完成的議題開始討論吧！
- Let's start by going over what we want to accomplish today.

- 大家都知道今天召開這個會議的目的吧！
- I think we all know why we're here today.

- 我們一定要先弄清楚今天討論的目的。
- We can't begin without first knowing what we hope to accomplish.

- 如果會議再不開始，明天又得從頭來過。
- If we can't get started, we'll have to do it all over again tomorrow.

- 會議拖得越久，我們就會越晚回家。
- The longer we put this off, the longer we'll have to stay here.

- 看來我得把菜刀架在你的脖子上才能開始討論。（玩笑口吻）
- If I have to light a fire under you to get this started, I will. [joking]

- 大家趕快就緒，開始開會吧！
- Let's get a move on and start the meeting already.

- 請問會議可以開始了嗎？
- Can we just hold this meeting, please?

- 我們可以開始了嗎？
- Can we just begin already?

- 各位，時間就是金錢。
- Time is money, people!

- 我的青春都耗費在等這個會議開始。
- I'm growing old waiting for you guys to settle down.

- 希望世紀末以前這個會議能夠正式開始。（諷刺語氣）
- I'd like to begin sometime during this century.[sarcasm]

輕鬆

028 如何結束會議

專業

- 有了大家的共識，我想今天的會議可以圓滿地告一個段落了。
  - With your concurrence, I think we're at a good point to adjourn for the day.

- 謝謝各位撥空前來參與這個重要的會議。
  - Thank you for coming and being a part of this important meeting.

- 相信大家都很滿意今天會議的成果。
  - I think we can all feel good about what we accomplished today.

- 我想今天的討論涵蓋了所有的議題，如果沒有異議，會議到此結束。
  - It seems that we've covered everything we needed to— let's call it a day, shall we?

- 期待我們下週／下個月／來年的聚會。
  - Let's come back to this when we reconvene next week/month/year.

- 今天在此告一個段落，希望大家回去能好好思考我們討論的內容。
  - Let's disperse and think about everything we've talked about.

- 今天的討論成果十分豐碩，謝謝大家。
  - Congratulations on a job well done, everyone.

- 我想，今天的會議非常成功。
  - Everything turned out quite well, I think.

- 我想今天的討論必須要先告一個段落。
  - I think we've done enough for today, don't you?

- 其他議題就留到下次再討論吧！
  - Let's table the other items until next time.

- 既然會議暫時沒有進展，今天先在此打住吧！
  - We aren't getting anywhere, so let's stop for today.

- 就算繼續討論下去，大家應該也已經不知所云了吧！
  - If this goes on much longer, we are all just going to mentally check out.

- 現在大家的討論都只是在原地打轉而已。
- These talks are no longer accomplishing anything.

- 今天就到此為止好嗎？
- Let's call it quits for today, okay?

- 我們該休息了。
- Let's give it a rest.

- 我沒辦法繼續討論下去了。
- I'm done with this meeting.

- 會議早該結束了。
- This meeting is so over!

- 不管，我要走了。
- I'm out of here!

- 我要去喝一杯，誰想加入？
- I feel happy hour coming on—who's with me?

無禮
- 現在應該是三更半夜了吧！
- It must be 5 o'clock somewhere!

029 如何讓談話重新聚焦

正面

- 我很歡迎大家提出新的點子，但請不要偏離主題。
- I always want to encourage new ideas, but let's stay on track.

- 你說得很對，不過我們還是先回到剛才的討論。
- That's very true, but let's get back to where we were.

- 你的發言我聽見了，但是目前應該先聚焦在今天的議題討論。
- I hear what you're saying, but for now let's keep on topic.

- 你的想法很好，但還是讓我們先把原本的議題討論完。
- I like the way you think, but let's continue where we left off.

- 大家先把注意力放在手頭這個案子上。
- Let's stay focused on the task at hand.

- 我想大家有點離題了。
- I think we're getting a little off track here.

- 希望你們還記得，今天會議的目的是…。
- As you may recall, the purpose of this meeting was...

- 我們別忘了今天討論的目的是…。
- We have to remember that the purpose of this meeting is...

- 今天所討論的重點是…。
- The essence of the topic is still...

- 我想我們應該回過頭去，把原本討論的議題先討論完。
- I think we should take a few steps back and finish the original discussion.

- 我們先不要偏離主題。
- Let's stay on track for now.

- 最糟糕的就是沒有討論到原本提出的重點。
- It's a shame if we let the essentials slip through our fingers.

- 大家不要一直鑽牛角尖。
- Let's not turn onto a dead end.

- 我們不要因為小事而模糊了焦點。
- Let's not get lost in trivialities/tangents.

- 我們已經越扯越遠了。
- We're not staying focused here.

- 討論離題對大家都沒好處。
- We would all benefit if we stayed focused.

- 請針對議題發言，拜託！
- Let's stay on point, please.

- 我想這整個討論已經失去了焦點。
- I think we're getting distracted from the real issue.

- 我們不應該這麼零零散散的討論。
- We don't need to take these talks in a different direction.

- 我們離主題越來越遠了。
- We're moving further and further off track.

- 大家已經徹底離題了，從頭開始吧！
- We've managed to completely go off topic; let's start over.

- 現在根本是在討論另一個新的議題吧！
- That is a whole topic unto itself.

- 我們完全沒有抓住一開始的重點。
- We are completely missing the point now.

- 我不想再繼續這種枝微末節的討論。
- I'm tired of all the tangents.

負面
- 這根本和議題八竿子打不著。
- That is totally besides the point!

030 如何提議一項行動或方案

溫和

- 我不確定可不可行，但如果我們這樣做呢？
  - This may not work, but what if we did X?

- 我的想法可能比較大膽，但我們何不試試…？
  - I could be way off, but what if we tried X?

- 如果可以的話，我建議…。
  - If I may, I'd like to propose that we do X.

- 我有個點子，想聽聽你的意見。
  - I'd like to get your thoughts on this proposal.

- 如果我們…，你覺得大家會怎麼想？
  - What would the group think if we did X?

- 我們可以討論一下…的優缺點嗎？
  - Can we discuss the pros and cons of X?

- 針對這件事，我想到了一個可能的解決辦法。
  - I've got a possible solution that may or may not fit the bill.

- 我可能已經找到這個問題的答案了。
  - I may have an answer to that question.

- 我們應該盡量集思廣益，而我自己想到的辦法是…。
  - We need to explore all avenues, but here's one idea...

- 條條大路通羅馬，我想到的點子是…。
  - There's more than one way to skin a cat; here's what I think...

- 這個問題有好幾種可能的答案，其中之一是…。
  - There are several possible answers to this; here is but one example...

- 或許我們可以從另一個角度來看。
  - Maybe we can look at the problem from a different perspective.

- 這個領域的專家們認為…。
  - Experts seem to think that...

- 在這種情況下，有人曾經…。
- Other people have done X in this situation.

- 我有個解決辦法，讓我解釋一下。
- I've got a way to move forward; let me explain.

- 我以前遇過類似的問題，我的建議是…。
- I've seen problems like this before, and I suggest...

- 我想我已經找到了問題的對策。
- I have arrived at what I believe is a workable solution.

- 這個問題非解決不可，我建議…。
- We need to find a solution to this mess; here is what I propose...

- 想了很久，我覺得最好的辦法是…。
- After much consideration, I believe the best course of action would be to...

- 我覺得接下來最合理的做法應該是…。
- I think we need to focus on the next logical step, which is...

- 我認為必須採取的行動是…。
- I submit that the following will be necessary.

- 我想已經到了該…的時候了。
- I think the time has come for us to do X.

- 除了…，我們沒有其他辦法。
- There's no way we can avoid doing X.

- 接下來，我們必須…。
- Here's what we're going to do.

- 這是唯一的辦法，就這樣。
- This is the only possible solution and that's final

- 我實在想不到其他更好的辦法了，你呢？
- I haven't heard a better solution yet, have you?

- 我們已經無計可施，這是唯一的路。
- There is no alternative but to do this.

強硬
- 你要嘛就加入我，不然就反對到底！
- Either you're with me or you're against me.

031 如何推延一項任務

禮貌

· 我也希望能處理這件事，但現在真的忙不過來。	· I wish I could address this, but I'm completely tapped out right now.
· 我現在忙得焦頭爛額，可以另外再找時間討論嗎？	· I'm swamped right now—can we revisit this some other time?
· 我很樂意下週／下個月／明年好好找個時間和你討論。	· I'd be delighted to set aside time to go over this [next week/month/year].
· 我很樂意在〔什麼時候〕和你討論這件事。	· It will be my pleasure to get back to you about that on [specific time].
· 如果你沒那麼急，我可以在〔什麼時候〕處理這件事。	· If you can hold off for a bit, I'll take care of it on [specific time].
· 我保證我一有空就會告訴你。	· I'll get back to it when I'm free I promise.
· 很抱歉，但是我現在真的沒那麼多時間處理這件事	· Sorry to put you off, but I just can't give this task the time it deserves right now.
· 目前我真的沒空，可以另外找時間處理嗎？	· I can't prioritize this right now. Is it okay that I take care of it some other time?
· 抱歉，我必須先暫停這件事。	· Unfortunately, I'll have to delay.
· 過些時候我可能有空處理這件事，但我不能打包票。	· I may have time to work on this later but I can't promise anything.
· 我知道這件事很重要，但我真的沒辦法優先處理。	· I understand that this is very important, but it's not my priority right now.
· 你晚點再過來吧！	· You'll have to come back some other time.
· 我們不如晚點再來談這件事。	· Why don't we let that go for now?
· 我們另外再找時間處理這件事吧！	· We'll deal with that some other time.

日常對話

辦公室厚黑學

②

處理需求與情緒

海外外交手腕

談判時刻

解決問題

表現風度體貼

掌控發言權的帝王學

- 等我有空自然會處理這件事。
- This will be dealt with when I have the time.

- 我的時間我自己管理。
- Well, you don't manage my calendar, do you?

- 我今天不能處理這件事，但我有空的時候會通知你。
- I'm not dealing with your request today, but I'll let you know if and when I do.

- 現在不是時候，你根本是在強人所難。
- You're being unreasonable—now is not the time.

- 我根本沒時間回應你的需求。
- I don't even have a second to entertain your request.

- 下禮拜再來找我看看。
- Ask me next week if I care.

直率

- 好好好，馬上去辦。（諷刺語氣）
- Yeah, that's gonna happen! [sarcasm]

032 如何推延一場對話

禮貌

- 我希望能夠另外在〔某時間點〕再來討論這件事。
- I'd be delighted to set aside time for this on [insert specific time].

- 我希望可以過幾天再來討論。
- I would prefer if we discussed this at a later date.

- 過些日子我們再撥多一點時間討論這件事。
- We'll cover more about this at a later time.

- 我們等到〔某時間點〕再來談這件事吧！
- Let's push this discussion until [insert specific time].

- 先讓我了解一下狀況，然後我們〔某時間點〕再來談。
- Let me get up to speed and we'll chat [insert specific time].

- 這件事需要謹慎處理，我們應該等到資訊更充足的時候再來討論。
- The question/issue deserves some serious thought—let's revisit this when we know more.

- 可以等到〔某時間點〕嗎？我希望能在比較有空／有精神的時候談這件事。
- I'd be more comfortable if we could talk when I'm less distracted/busy/tired—how about [insert specific time]?

- 我會發一封郵件給你詳細討論這件事。
- I'll send you an e-mail and address that in greater depth.

- 我得先思考一下才能和你討論這件事。
- I'm going to need to think about this and get back to you.

- 和你談之前，我需要先整理一下這整件事。
- I need to go over this in my head first.

- 我必須要多做一點研究才能和你談論這件事。
- I will need to learn more about the topic before we talk.

- 必要的時候我會找你談，現在還太早了。
- We'll talk about it when it becomes necessary, but not before.

- 我晚點回覆你。
- Let me get back to you on that.

- 先讓我想一下。
- Let me think about it.

- 我們把討論延到〔某時間點〕好嗎？
- Let's push this discussion to [insert specific time/date].

- 我想現在不是談這件事情的時候。
- I don't think this is the best time to talk about this.

- 目前沒什麼好說的，但我相信過些時候我們可以好好談談這件事。
- I have nothing to add right now—we'll talk about it again, I'm sure.

- 我們得另外找時間談。
- We'll have to talk some other time.

- 我們過些時候再來談這件事情，好嗎？
- Let's revisit this some other time, okay?

- 我們一定要現在談這件事情嗎？
- Do we have to talk about this right now?

- 我今天不想再繼續討論這件事情了。
- Continuing this conversation today is out of the question.

- 我剛剛已經說了，找時間再說。
- I said we'd do it another time.

- 我現在沒辦法跟你談。
- I can't handle this right now.

- 我現在根本沒辦法聽你講話。
- There's no way I can listen to you now.

- 你怎麼會覺得我會在這個時候和你談這件事？
- You're crazy if you think I'm going to talk to you about this now.

粗魯

- 別再煩我了，等我準備好自然會去找你。
- Stop bothering me—I'll talk about it when I'm good and ready.

033 如何推延做決定的時間

禮貌

- 不要這麼衝動，我們再好好想一下。
- Only fools rush in—let's think on this some more.

- 與其做出錯的決定，不如花多點時間找出正確的對策。
- Let's take our time to find the right solution rather than rush to a mistake.

- 我們應該先花點時間好好想想我們要的是什麼。
- Let's take some time and think about what we want to accomplish here.

- 我們沒必要現在做決定。
- We don't need to make a decision right this very moment.

- 不如先暫停一下吧。
- Let's sleep on it first, okay?

- 做出最後決定前我們應該先冷靜一下。
- I think we should get some distance on this before we decide.

- 我們先把這件事放一邊，晚點再說。
- Let's table this until later.

- 我想這件事可以晚點再決定。
- I think we can make this judgment later.

- 我保證會好好考慮以後再給你答覆。
- I'll definitely think about it and get back to you.

- 你什麼時候需要答案？
- When do you need an answer to this?

- 等我有了答案再告訴你。
- Let me get back to you with my verdict.

- 我們可以過陣子再來討論這件事。
- We can come back to this issue in the future.

- 在做出決定之前，我們必須更進一步研究。
- This decision may require further analysis.

- 我們先思考真正的目的是什麼，再回來討論這件事。
- Let's think about what we really want and get back to it later.

- 現在只是在原地打轉，晚點再說吧！
- We're not making any progress—let's talk about it later.

- 我們應該等一切都塵埃落定後再來討論。
- We should work on this when the dust has settled.

- 我不認為現在是討論這件事的好時機。
- I don't think now is the time discuss this.

- 我想先冷靜一下對大家都好。
- I think we could all benefit from a cooling off period.

- 我們晚點再討論。
- We will discuss this later.

- 等時機更適當的時候再來談吧！
- We'll talk about it when the time comes.

- 我不會因為時間壓力而草率決定。
- I will not bow to the tyranny of the urgent.

粗魯
- 我目前沒有答案，不要再問了。
- I'm not going to give my decision right now, and that's final!

日常對話

辦公室厚黑學 ②

處理衝突與情緒

巧環外交手腕

臨判時刻

解決問題

表現風度禮貌

掌控發言權的帝王學

034 如何團結眾人

鼓勵

· 唯有彼此合作才能夠成功。	· Only by working with one another will we succeed.
· 如果大家心中都有一樣的願景，沒有什麼事能難倒我們。	· We'll be unstoppable if we have a common vision.
· 只要我們大家同心協力，再大的難關也可以度過。	· We can overcome any obstacle if we work together toward a common goal.
· 認為自己是團隊一份子的人請舉手。	· Raise your hand if you're a team player!
· 只要我們目標一致，再大的困難也不怕。	· Once we agree, nothing will stop us.
· 大家目標一致才能有高昂的士氣。	· We're all happiest when our goals are one and the same.
· 唯有攜手合作，才能一起創造明天的歷史。	· Tomorrow's page—no one can write it alone.
· 團結力量大。	· Together, we are stronger than we are as individuals.
· 三個臭皮匠，勝過一個諸葛亮。	· A strand of three [or 20, or 200] is not easily broken.
· 我們的向心力遠遠大過彼此的歧見。	· The bonds that hold us are stronger than the forces that separate us.
· 未來的成敗就看我們這一次怎麼表現了。	· How we deal with these changes now will make or break our future.
· 讓我們攜手合作，一同前進。	· Let's work together and move forward.
· 我想大家都同意…。	· I think we can all agree that...

- 想要獲得更大的成就，我們就必須團結一致。
- Together, we can take the high road and succeed.

- 我希望你能加入我們，一起解決問題。
- I'd like to see you join with me in solving this issue.

- 大家把眼光放遠一點。
- Let's all look at the big picture.

- 讓我們一鼓作氣，向前邁進。
- Let's proceed in a spirit of togetherness.

- 公司的成長，是每個人共同的目標。
- The good of the company should be our common goal.

- 只要我們團結，一定會成功。
- We will succeed if we work as one unit.

- 我知道大家都願意互相合作，現在正是表現的時候。
- I know you all have the capacity to work together, but now you need to show me.

- 越多人一起加入，就能越快完成。
- We'll get this done faster if we all work together.

- 大家的目標一致，才能夠克服種種困難。
- A common goal will help head off problems down the road.

- 團結而存，分裂必亡。
- United we stand, divided we fall.

- 內鬥爭權不會有好下場。
- Infighting and power plays will get you nowhere.

- 這裡不歡迎三心二意、朝秦暮楚的人。
- This is no place for fence-sitters or the partially committed.

- 在這裡，大我勝過小我。
- The collective takes precedence over the individual here.

- 如果你不和我合作，那麼請你離開。
- We either stay together, or you get out of the way.

威脅

- 我們不接受任何異議份子。
- We don't tolerate dissent within our ranks.

035 如何稱讚上司

誇張

- 我真的對你佩服得五體投地！
- I worship the ground you walk on—is that wrong?

- 你是我的偶像。
- I put you up on a pedestal.

- 這絕對是我看過最好的報告／簡報／分析。
- This is easily the best [report/briefing/analysis/work] that I've ever seen.

- 你…的手法真是太令人佩服了。
- The way you do X is simply amazing!

- 我打從心裡欣賞你處理每件事的手腕。
- I'm amazed at how you handle everything.

- 看你做事簡直是種享受。
- I love watching you in action.

- 我每天都從你身上學到許多東西。
- I learn so much from you every day.

- 你的工作能力總是這麼傑出。
- Your work sets you apart from everyone else.

- 你的毅力／道德／決心深深鼓舞著我。
- I am inspired by your determination/work ethic/will to succeed.

- 身為你團隊的一份子，我覺得很榮幸。
- I'm proud to be on your team/working for you.

- 很少有人像你這樣兼具能力與品德。
- Men/women of talent and integrity are rare.

- 我不該這麼不斷地讚美你，但我就是忍不住。
- I shouldn't go on so much about your work, but I can't help myself.

- 我相信我們將會有一段美好的合作關係。
- This could be the beginning of a beautiful partnership.

- 以你的工作表現，一定可以步步高升。
- With results like these, you'll be unstoppable.

- 你認真工作的態度，絕不會一直困在這個小池塘裡的。（玩笑口吻）
- With your discipline, you won't be staying in the mail room forever. [joking]

- 我很高興能加入你的團隊。
- It's good to be on board with you and your team.

- 如果我像你這麼認真，應該早就升到副總了。
- If I worked as hard as you, I would have made partner by now.

- 你的成就說明了你對工作投入多大的心血。
- Your achievements speak volumes about your dedication

- 所有的讚賞肯定，你都當之無愧。
- You deserve every accolade you receive.

- 每個人都應該學習你認真工作的精神。
- Everybody should put as much gusto in their work as you do.

- 如果大家都像你這麼努力，公司業績一定一飛沖天。
- If everyone worked as hard as you, this company would be ahead of the game.

- 你搶了我們所有人的鋒頭。（玩笑口吻）
- Once again, you make us all look bad. [joking]

- 你的表現總是讓大家驚艷。
- Nice work, as usual.

- 嘿，還不錯嘛！
- Not too shabby, partner!

036 如何鼓舞員工

正面

- 你是個很棒的工作夥伴，請保持下去。
- You are wonderful to work with—keep it up!

- 我可以看到你離加薪／紅利／升遷不遠了。
- I can see your bonus/raise/promotion from here!

- 繼續維持你的高水準表現，大家一定會注意到你的。
- Keep up the great work—it won't go unnoticed.

- 千萬、絕對，不要放棄。
- Never, ever, ever give up!

- 現在正是你加碼衝刺的最佳時機。
- There's no time like the present to kick it into high gear.

- 我會在這裡支持鼓勵你往前邁進。
- I'm right behind you, encouraging you with each step forward.

- 只差一步你就抵達終點了。
- You are so close to the finish line!

- 我以你為榮，希望你繼續在工作上有出色的表現。
- I am proud of you—keep up the good work.

- 只要有你在我的團隊，最後總是能夠達成任務。
- When we work as a team we always get great results.

- 現在不是鬆懈的時候，我們一起堅持到最後一刻吧！
- Now isn't the time to stop—let's press on to the end.

- 只要你用心，一定可以脫穎而出。
- You can become one of the elite if you put your mind to it.

- 只要撐住，眼前的難關總是會過去的。
- Hard times will soon be a thing of the past if you hang in there.

- 千里之行，始於足下。
- Even the longest journey begins with a single step.

- 記住你真正想要達成的目標。
- Keep your eyes on the prize.

- 最終回到一個基本的問題：你想要創造什麼樣的未來？
  - It comes down to a single question: What future do you want to create?

- 你的價值就看你自己的決定了。
  - You are the employee you decide to become.

- 一切看你對這份工作做出的承諾有多大。
  - Just how committed are you to making this job work?

- 你一定要保持專注。
  - You need to keep your nose to the grindstone

- 公司的未來在你手裡，別搞砸了。
  - Our company's future is in your hands—don't drop the ball.

- 我已經看到一絲曙光，別擔心。（玩笑口吻）
  - I can see the light at the end of the tunnel, and it isn't an oncoming train. [joking]

- 我們的確是有往前邁進，但還沒有抵達終點。
  - We're making some progress, but we're still not there yet.

- 你不一定要延長工作時間，但是要改進工作效率。
  - You need to work smarter, not necessarily harder.

- 不盡全力的話是沒辦法完成這項工作的。
  - The job won't get done if you don't pull your weight.

- 現在已經沒有退路了，只能繼續向前。
  - The only place to go from here is up.

- 公司期待你有更好的表現，希望你已經準備好接受挑戰。
  - The company is expecting more from you—I hope you're up for it.

- 我相信你下次會更進步。
  - I'm sure you will do better next time.

- 你的一舉一動我們都在注意。
  - We are watching you every moment of every day.

- 我不想再重複講一樣的話！我要看到績效。
  - I don't want to hear myself talking; I want progress.

- 如果達不到標準，你就得走人。
  - You either cut the mustard or you're done.

負面

037 當員工表現退步時

溫和

- 你的工作量太重了嗎？也許我可以幫忙。
  - Is your work load too stressful? Maybe I can help.

- 你是不是有什麼心事？需要聊一下嗎？
  - Is there something on your mind? Something you'd like to discuss?

- 你看起來心事重重／心不在焉？有什麼煩惱嗎？
  - Are you having any issues away from work? You seem distracted/unhappy/disengaged.

- 最近你的工作表現退步，有什麼方法可以改善？
  - Your work quality has been suffering as of late—what can we do to turn it around?

- 或許你需要休個假，調整一下你的腳步。
  - Perhaps you need a little break to regroup.

- 我知道你的能力不只是這樣而已。
  - I know you are capable of much more than this.

- 你一定要做好你的工作才能在團隊裡站穩腳步。
  - You need to carry your weight in order to get the recognition you deserve.

- 你明明有很好的資質，為什麼不好好表現呢？
  - You have such potential—why are failing to follow through?

- 我知道你的能力遠遠超出你最近的表現。
  - I know you've got a lot more talent than what I've been seeing lately.

- 這種令人失望的表現一點也不像你。
  - This kind of underwhelming performance isn't like you.

- 你最近的表現完全在水準之下，到底怎麼回事？
  - Your performance has been substandard lately—how come?

- 現在應該是你展現真正能力的時候，不是嗎？
  - Isn't it time that you showed us what you're capable of?

- 大家都想要幫你，你願意接受嗎？
- Team members have offered to help you—is that what you want?

- 這究竟是怎麼回事？我甚至必須要把你的工作交代給其他人！
- I've had to delegate your tasks to other people—why is that?

- 公司覺得你應該要有更好的表現。
- The company is expecting a lot more from you.

- 在這間公司，每個人都要為自己的工作表現負責。
- In this company, we take responsibility for our conduct.

- 你最近是怎麼了？
- What's gotten into you lately?

- 整個部門都在密切觀察你的工作表現。
- Our department head is looking at your performance closely.

- 我認為你應該更用心在工作上。
- Taking on more responsibility at work might be a good idea.

- 我們都期待你能有更好的表現，而且大家都在觀察中。
- We all expect a lot from you, and we're watching.

- 你拖累了所有人的進度，不能再這樣下去了。
- You're holding everyone back—this can't go on forever.

- 這種半吊子的工作表現絕對會引來上面不滿。
- This kind of shoddy performance warrants a verbal warning.

- 我已經聽到大家對你的批評，你最好注意一點。
- I've heard people whispering about you; you'd better get it under control.

- 我會馬上向上面檢舉你。
- I am going to have to write you up immediately.

- 沒有表現就沒有薪水。
- You need to earn your pay.

- 如果你跟不上公司的腳步，只能請你另謀高就。
- If you can't pick up the pace, we'll have to let you go.

- 如果你到〔某時間點〕還沒有進步，就請你離開。
- If you don't get yourself together by [specific date], you will be terminated.

嚴厲

99

038 如何開除員工

專業

- 很抱歉，我們已經沒辦法繼續聘請你。
- We can no longer afford to keep you on, unfortunately.

- 恐怕我們已經無法繼續聘用你了。
- I'm afraid it's just not working out any longer.

- 很抱歉，我們必須終止和你的合作關係。
- I'm sorry, but we're going to have to let you go.

- 我相信還有其他更適合你的地方。
- I know you will be much happier elsewhere.

- 你是個聰明的人，一定會找到自己出路的。
- You're a smart person—we all know you'll land on your feet.

- 如果你能投入更多心力，有一天一定會成為一個優秀的員工。
- Someday you will make an excellent employee if you put your mind to it.

- 我們需要針對你的工作表現好好談談。
- We need to have a serious discussion about your work performance.

- 今天是你在公司的最後一天。
- This is going to be your last day with the company.

- 我們沒有其他辦法，只能請你離開。
- We have no choice but to let you go.

- 公司已經嘗試各種方法提升你的工作表現，但還是不符合我們的要求。
- We tried our best to help you but we still didn't get the required work quality.

- 你的行為完全不符合公司規範，我們必須請你離開。
- Your behavior flies in the face of SOP; we have to let you go.

- 如果繼續容忍你的錯誤，可能連我都會丟了工作。
- I cannot risk losing my job for your mistakes.

- 你犯的錯誤無法原諒，我們必須請你離開公司。
- We can't condone what you did; I have no choice but to let you go with cause.

- 你違反了規定，自然需要付出代價。
- You broke the rules and now you have to pay the price.

- 你的表現讓我除了開除你，沒有其他辦法。
- You leave me no choice but to fire you.

- 希望你明白，這完全是公事公辦，和你個人無關。
- You know, it's nothing personal—it's just work.

- 你覺得到了這個關頭會發生什麼事呢？
- What do you think should happen to you at this point?

- 我收到許多對你的投訴，所以只能請你走路了。
- I have a stack of complaints against you; there is nothing left for me to do but fire you.

- 你真的不適合公司，我必須請你離職。
- You're clearly not a good fit for this company; I have to let you go.

- 你能撐到現在已經出乎我意料了。
- I'm surprised you made it this far.

- 我們都很欣賞你的人格，但無法接受你這麼缺乏紀律／動力。
- We'll miss your personality, but not your lack of discipline/motivation/dedication.

- 你能在公司做到今天已經是很離譜了。
- It's a shame that our code of conduct allowed you to be here this long.

- 我已經對你非常忍耐了。
- I've been more than patient with you.

- 抱歉，你被開除了。
- Sorry, but you're out.

- 我正式宣布，你被開除了。
- It's official—you're fired.

- 公司很早就已經決定要開除你了。
- Firing you has been a long time coming.

- 公司正在轉型，你已經不再適任。
- We're going in a different direction, and you won't be along for the ride.

- 請收拾好東西自行離開。
- Feel free to show yourself out.

- 等下離開公司的時候，小心慢走。
- Don't let the door hit you on your way out.

- 你大概沒辦法在這個產業繼續混下去了。
- You're probably finished in this business.

- 我要看著你離開！
- I'm going to enjoy watching you leave.

- 你已經是過去式了！
- You're history.

不專業
- 滾出這間公司！
- You're outta here!

039 如何強調事情的急迫性

明顯

・加把勁！ ・ Step on it!

・該行動了！ ・ It's go time!

・各位，我們開始吧！ ・ Let's get a move on it, people!

・我們沒時間可以浪費了。 ・ We don't have a second to waste.

・時間緊迫。 ・ Time is of the essence.

・你還在等什麼？時間寶貴。 ・ What are you waiting for? Time is a-wasting!

・現在已經沒有退路，我們趕快開始吧！ ・ Failure is not an option, so let's get going!

・你知道什麼叫做最後關頭嗎？就是現在！ ・ Have you heard of the last minute? Well, this is it!

・各位，時間是不等人的。 ・ The clock is ticking, folks.

・拜託大家專心點，時間已經不多了。 ・ Please, let's focus. We only have so much time to finish.

・我需要你們的幫忙，真的很緊急。 ・ Please understand the urgency—I need your help now.

・如果我們能專心並且掌握好時間，應該可以如期完成。 ・ We can get this done if we focus and keep track of time.

・現在去做就對了。 ・ There is no later; there is only now.

・我們不能再浪費時間了。 ・ We need to stop wasting time.

・我們已經沒時間繼續耽擱下去了。 ・ We no longer have the luxury of time to put this off.

- 過去讓它過去了，重點是把握現在。
- We need to put our wasted time behind us and keep going.

- 我們要好好把握剩下不多的寶貴時間。
- We need to use what little time we have left.

- 事情已經迫在眉睫。
- Time is a luxury we no longer have.

- 今日事，今日畢。
- Why put off 'til tomorrow what you can do today?

- 事情很多，時間很少。
- So much work, so little time.

含蓄

- 現在可不是你曬太陽的時候。（諷刺語氣）
- This isn't the time to be working on your tan. [sarcastic]

040 如何緩和事情的節奏

風度

- 我想我們應該在做決定之前再檢視一次所有的方案。
- I think we need to ponder all our options before making a decision.

- 這麼重要的決定應該要花更多的時間及心力。
- A decision this important should be given the time and attention it deserves.

- 我很欣賞你的積極，但現在應該稍微放慢腳步。
- I appreciate your enthusiasm, but let's slow down for a moment.

- 我很欣賞你的效率，但我們不應該貿然行動。
- I appreciate your alacrity, but we shouldn't rush this.

- 小心起見，這件事我們應該慢慢進行。
- For the sake of thoroughness, I think we should take our time.

- 不如我們多花點時間從長計議吧！
- What if we took some time to cogitate on this?

- 事情進行地這麼快不一定對我們有利。
- It might be to our disadvantage to work this quickly.

- 強摘的瓜不甜。（玩笑語氣）
- We sell no wine before it's time. [joking]

- 我們應該放慢腳步，先好好瞭解事情的始末。
- Can we delay for the purpose of understanding better?

- 我們先暫停一下，把整件事好好想一遍。
- Let's slow down and really think this through.

- 時間還很多，我們不如慢慢來。
- Let's take baby steps while there's still so much time.

- 我們不要為求快而把事情搞砸了。
- Let's not burn any bridges by moving too quickly.

- 現在還不到做決定的時候。
- It's early yet in the decision-making process.

- 別慌，時間還很充裕。
- There's plenty of time; no need to panic.

- 我們不應該急病亂投醫。
- We shouldn't rush into things.

- 每一件事都依照時間表進行中。
- Everything in good time, my friend.

- 匆促之間下的決定通常都不會有好結果。
- Snap decisions rarely work out well in the end.

- 何必著急呢？時間還有很多啊！
- Why rush? We've got more than enough time.

- 我想現在做決定有點言之過早。
- I think that it is a bit premature to make a decision.

- 照這樣的速度進行下去，很容易發生問題。
- The speed at which we're working is creating a space for error and confusion.

- 倉促之下犯的錯誤，要花很多工夫修補。
- Make an error in haste, repent at leisure.

- 聰明的人懂得三思而後行。
- Only fools rush in.

- 可以把速度放慢點嗎？我需要喘口氣。
- Can we slow down a bit? I need to catch my breath!

- 不需要驚慌失措，我們還有時間。
- We're in no hurry—no need to cause a panic!

- 這樣的速度簡直是瘋狂。
- It's madness to move at this speed.

- 忙中必定有錯。
- Haste makes waste.

- 一步一步走穩再說。
- One step at a time.

- 你的鹵莽行事會對我們造成困擾。
- Your impulsiveness is not serving us well here.

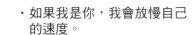

- 如果我是你，我會放慢自己的速度。
- If I were you, I'd slow down a bit.

- 可以用走的，你又何必跑呢？
- Why run when you can walk?

- 沒有人在倒數計時，你不用這麼急。
- No one here is watching the clock, so slow down.

- 好好想想再做吧！
- Think, think again, then act!

- 我們該踩煞車了。
- Let's put the brakes on, okay?

- 嘿，慢點吧！你這個急驚風。
- Whoa, slow down there, speedy!

無禮

- 衝這麼快，是哪裡失火了嗎？
- Where's the fire?

日常對話

辦公室厚黑學

2

處理衝突與嫌隙

跟他外交辭令

談判時刻

解決問題

表現風度翩翩

掌控發言權的帝王學

Part

處理衝突與憤怒

人與人相處,難免會有摩擦、甚至衝突的時候,把握住以下四個原則:溝通與妥協、就事論事、態度冷靜頭腦清晰、同理心,應可順利通過考驗。

溝通哪些事

- 如何化解緊張氣氛
- 如何調解其他人的衝突
- 如何化解誤會
- 如何回應充滿敵意的評論
- 當對方生氣時
- 被緊迫盯人時
- 當他人固執己見
- 當他人態度傲慢
- 當他人出言威脅
- 當有人對你百般挑剔
- 當有人說謊
- 當有人想和你吵架

- 當被人打斷時
- 當被人取笑時
- 當被人批評時
- 當他人防衛心過高
- 當被人質疑時
- 當有人逃避你的問題
- 當有人對你吼叫
- 當有人對你罵髒話
- 當有人想轉移話題
- 當被人侮辱時
- 當有人對你暴力相向

> *無法遏制怒氣的人總是做出錯誤的決定。*
> ——威爾・羅傑斯（*Will Rogers*, 美國三〇年代演員）

　　大多數的人際關係裡總是難免會產生衝突，下面幾個原則能夠讓你通過衝突的考驗，甚至和對方建立起更穩固的連結：

 溝通與妥協

　　要擁有一段健康的關係需要具備兩種能力：坦誠溝通和相互妥協。這兩種能力可幫助你有效化解衝突。我們不能一昧地避免衝突，反而應該在控制好情緒的情況下和對方開誠布公地討論。請務必記得，溝通的目的是解決問題，而不是贏得勝利。如果你永遠要做「對的一方」，那麼不管在家裡或職場，都很難擁有正常健康的人際關係。

② 就事論事

　　很多時候衝突無法解決的原因是因為討論的內容已經脫離主題或者是鑽牛角尖。在還沒得出結論之前，記得就事論事。不管是否相關，別翻舊帳，那只會使衝突程度更加嚴重。離題的討論只會耗損大家的能量。所以，保持耐心、集中焦點，問題通常會很快順利解決。

日常對話

辦公室裡眉角

處理衝突與憤怒 ③

展現外交手腕

談判陣術

解決問題

表現風度體貼

掌控發言權的帝王學

❸ 態度冷靜和頭腦清晰

冷靜的態度和清晰的思路永遠是跨越衝突的不二法門。記得要保持正面的態度，不要大吼大叫、指名道姓，或是人身攻擊。如果當下真的無法克制怒氣，不妨給自己一些時間冷靜下來再處理。當你遠離衝突場面的時候，試著建構出可能的正面對策，然後沙盤推演該如何將討論導向希望的結果。在自我冷靜的期間，請保持正面思考，也別過度解讀問題。

❹ 抱著同理心聆聽

提到衝突場面的溝通技巧，很多人都忽略了讓對方把話講完這個簡單的動作。不管你有多想要插話，千萬不要打斷對方，專心地聽，並且保持眼神接觸讓對方知道你正在聆聽。對方說完之後，試著站在對方的角度去理解，然後再做出反應。一旦你的善意被察覺，氣氛自然會緩和下來。請記得，衝突管理的重點不在於分出誰輸誰贏，而是互相妥協，試著找出大家都滿意的解決辦法。

041 如何化解緊張氣氛

溫和

- 沒關係，我們只是看問題的角度不同罷了！
- It's okay, we just don't see eye to eye on this.

- 抱歉，我想我們只是不太明白彼此的意思。
- I'm sorry, I guess we just don't understand each other.

- 大家先把有爭議的話題擱在一邊比較好。
- I think it's generally better to stay away from controversial topics.

- 就當作我們昨晚沒睡飽／星期一症候群吧！（玩笑口吻）
- Let's just chalk it up to semantics/Mondays/not getting enough sleep. [joking]

- 我想就算是比我們聰明的人也未必能處理好這件事。
- This topic has caused a lot of trouble for people smarter than we are.

- 我們兩個人對這件事的了解似乎都不夠充分。
- Neither one of us knows enough to have a productive discussion.

- 我想我們應該在吵起來之前先暫停討論。
- Well, we should probably stop now before things get too heated.

- 不需要為了這一點小事情起爭執。
- There's no reason to fight over little things.

- 你不覺得這個問題非常的棘手嗎？
- Don't you feel how problematic this topic is.

- 你覺得我們還應該繼續討論下去嗎？
- Is it really wise to pursue this discussion?

- 這類問題往往造成不少爭議。
- Problems such as these tend to create a lot of controversy.

- 我們兩個不該為了這種小事吵架的。
- Let's not allow relatively minor things to put us in a state of conflict.

- 很多人都栽在這個問題上頭。
- This topic has caused a lot of trouble for a lot people.

- 我們只是看法不同，如此而已。
- It's just a difference of opinion—nothing more, nothing less.

- 我想大家態度冷靜一點對彼此都好。
- I think we would all benefit from a less hostile dynamic here.

- 不如我們各自去洗把臉，冷靜冷靜再回來討論。
- Let's take a breather and regroup when we're both calmer.

- 我們何必為了自己根本不了解的事爭得面紅耳赤呢？
- Why are we arguing when we actually know so little about this?

- 這種爭吵完全無濟於事。
- It's not constructive to have so much dissention.

- 我們不應該老是批判別人，應該更有修養的處理這件事。
- We shouldn't judge one another—let's be more civil.

- 我們或許把事情想得太過嚴重了。
- Aren't we making a big deal out of nothing?

- 我們應該對彼此多點尊重，氣氛會好多了。
- I don't see respect being offered from either side—let's start over.

- 你可能還想爭論下去，但是我真的不想吵架。
- You may want to argue, but I'm not the fighting kind.

- 如果你想找人吵架，恕我無法奉陪。
- If it's an argument you want, I'm not the person for the job.

- 在場面變得難看之前，我們先打住吧！
- Let's put a stop to this before it gets any worse.

- 再講下去，一不小心就會吵起來了。
- It wouldn't take much to make this misunderstanding a lot worse.

- 這些難聽的話其實對溝通一點幫助也沒有。
- All of this bickering isn't getting us anywhere.

- 與其這樣針鋒相對一整天，倒不如坐下來好好講。
- We could wrangle all day, but I'd rather work things out more peacefully.

- 我們等大家都冷靜點再繼續討論。
- We'll continue when everyone is acting less aggressively.

- 你和我說話的口氣可以不要那麼衝嗎？
- Would you be less combative when you speak to me, please?

- 你為什麼這麼生氣／憤怒／激動？
- Why are you getting so excited/upset/irascible?

- 你覺得你的反應對解決問題有幫助嗎？
- Do you think your behavior is bringing us any closer to a resolution?

- 不要妄下斷言！
- Don't jump to conclusions.

- 不要小題大作！
- Don't bicker about trivial matters.

- 我們不用為了這種小事情開戰吧！
- We don't need a war to figure this out.

- 請你冷靜。
- Cool your jets.

強烈
- 放輕鬆！
- Chill out!

如何調解其他人的衝突

 圓滑

- 這種情況下難免會生氣，像我有一次也是…。
- It's easy to get upset about this; why, just the other day this happened to me...

- 大家表現得文明點嘛。（玩笑語氣）
- Let's keep this sophisticated, people! [joking]

- 我應該請你們先簽下停火協定。（玩笑語氣）
- I demand a cease fire! [joking]

- 你們最好停下來，不然我就要請你們兩個走人了。（玩笑語氣）
- Calm down or I'll give you both a pink slip. [joking]

- 如果大家不要這麼情緒化，相信會是很棒的討論。
- This discussion could be interesting if everyone was less emotional.

- 在你們吵起來之前，我們先把今天的重點再看過一次。
- Before you start arguing, let's go over all the points again.

- 等一下，我想先講一下自己的一些淺見。
- Just a moment—I have something to say that might be useful.

- 拜託，我們應該要更有團隊精神才對。
- Come on, guys—we need to work as a team here.

- 大家都應該要維持和諧的氣氛。
- We want to maintain harmony in this environment.

- 我們應該是要彼此合作，而不是互相攻擊。
- We are here to collaborate, not bring each other down.

- 我們先暫停討論，把這件事放在一邊吧！
- What if we stopped this discussion and simply let it go for now?

- 大家先各自休息休息，等冷靜一點再回來討論。
- Let's take a breather and reconvene when everyone's calmer.

- 大家不要為了這些事而傷了彼此的和氣。
- Let's not kill the mood with those topics.

- 如果你們再這樣吵下去，最後一定什麼事都做不成。
- The project will be in danger if you keep fighting.

- 我們應該要互相扶持才是。
- We all need to be cordial with one another.

- 你們再不冷靜下來，所有人都會被拖下水的。
- You guys need to keep cool before we all get into trouble.

- 大家有話好好說。
- Let's discuss the subject in peace.

- 現在吵架根本無濟於事。
- It's pointless to argue at this time.

- 我們不應該把時間浪費在吵架上。
- We've got better things to do than yell at each other.

- 我們應該做建設性討論，爭吵是沒有意義的。
- Let's stay constructive, here; arguing is pointless.

- 大家就吵到這裡為止，好嗎？
- Let's not speak about it any longer, okay?

- 這樣吵下去什麼事也辦不成。
- You're not getting anything accomplished this way.

- 你們最好在說錯話之前停止吵架。
- Well, you should probably stop while you're ahead/before you say something you'll regret.

- 爭執是我們現在最應該避免的行為。
- That's the kind of behavior we should avoid right now.

- 我真的很不欣賞你們現在的態度，希望大家到此為止。
- I really dislike this kind of attitude and would like it to stop.

- 這應該是成年人的理性討論，不要像小孩子一樣吵吵鬧鬧的。
- This is a discussion between equals, not children engaged in a schoolyard fight.

- 如果你們再不停止，我就通知人事主管／你的老闆／你的父母。
- If you two don't knock it off, I'm going to have to speak with HR/the boss/your father/your mother.

- 如果你們不能找出解決方法，你們兩個都走人吧！
- If you can't settle down, you're both out of here.

- 給我冷靜下來！
- Hey, take a chill pill!

- 我覺得自己好像來到了精神病院。（諷刺語氣）
- It looks like the lunatics are in charge of this asylum.[sarcasm]

直接

043 如何化解誤會

技巧

- 我很看重我們的關係，所以我們彼此公平競爭好嗎？
- I value our relationship—let's play fair with one another, okay?

- 或許你不太樂意，但我想我們應該把細節再討論一次。
- I know this is tough—maybe we can both go over the details once more.

- 我想我們可以一起合作，化解過去的誤會。
- I believe we can work collaboratively and solve this misunderstanding.

- 過去我們都誤解了對方，我希望這次能澄清誤會。
- We're misunderstanding each other and I want to make things better.

- 或許我們都應該站在對方的立場想想。
- Searching for common ground would help keep us on track, here.

- 三個諸葛亮，勝過一個臭皮匠。我們再試一試吧！
- Two heads think better than one—let's give this another try.

- 做生意／家庭裡／朋友間難免會有問題需要解決。
- Solving problems is a part of business/family life/friendship.

- 我們彼此都得讓步才能解決問題。
- We both need to compromise a little to make it work.

- 如果你願意，我們可以重頭來過。
- We can rewind to the beginning and start over if you'd like.

- 把眼光放遠一點對我們兩個都有好處。
- It would help us both to focus on the larger picture.

- 希望你也同意我們先把彼此的誤會／衝突放在一邊。
- Please help me put this problem/issue/conflict to rest.

- 我想你有你的道理，可以解釋給我聽嗎？
- I'm sure this makes sense to you—can you explain it to me a bit better?

- 我知道你在想什麼，但是我和你的看法恐怕不一樣。
- I see where you're going with this, but I don't think we're on the same wavelength yet.

- 如果我們能夠保持客觀冷靜，對事情應該會有很大的幫助。
- I believe we'll both get further along if we keep our emotions out of it.

- 我不想再繼續爭執下去了，我們一次把事情講清楚吧！
- I don't want to clash anymore—let's figure this out once and for all.

- 你必須放下敵意，我們才能解決問題。
- You need to be less hostile so we can solve this problem.

- 只要你不再妄自揣測，我們的進度一定突飛猛進。
- Once you stop making erroneous assumptions, we'll be able to make real progress.

- 你為什麼不肯接納別人的意見呢？
- Why are you so closed off to the ideas of others?

- 拜託，我們別吵了。
- C'mon, let's not fight.

直接
- 哇，我早就忘了這件事了。
- Wow, let it go already!

如何回應充滿敵意的評論

友善

- 你說話可真是引經據典啊！（玩笑口吻）
- You certainly have a great command of the idiom! [joking]

- 哇，真不敢相信你居然會說出剛才那些話。（微笑貌）
- Wow, I can't believe you just said that! [smiling]

- 我沒有要和你吵架的意思，不過我很訝異你會說出這些話來。
- I'm not going to argue with you, but I'm still a bit surprised you said that.

- 我不認為大部分人會同意你的看法。
- I'm not sure that many people would agree with you there.

- 你不覺得有些事情還是不要討論比較好嗎？
- There are probably some topics that are best avoided, don't you think?

- 我不知道該不該把你的話當真。
- I don't know whether to take you seriously or not.

- 我很驚訝你會這麼說，真令我傷心。
- Your statements really upset me—I'm surprised you said that.

- 你的話讓我感到不安／根本人身攻擊／很不得體／令人困擾。
- I find that quite disturbing/offensive/tactless/upsetting.

- 你真的這麼想嗎？
- Is that really what you think?

- 你知道自己在說些什麼嗎？
- Do you realize what you're saying?

- 你應該知道自己的言論很莫名其妙吧？
- You do know that's completely unreasonable, right?

- 我沒有聽錯吧？
- Am I hearing you correctly?

- 我不欣賞你剛剛的發言。
- I don't appreciate your choice of words.

- 我不喜歡你剛剛說的內容。
- I don't like what you're saying very much.

119

- 你說這些話的意義在哪裡？
- What's the point of saying something like that?

- 你應該三思而後言。
- Don't let your words get ahead of your thoughts.

- 你不覺得你應該換個方式說嗎？
- I think you should rephrase that, don't you?

- 你真的了解自己剛剛說了些什麼嗎？
- I'm not sure you're thinking clearly right now.

- 我不敢相信你竟然這麼說／支持／反對這件事。
- I can't believe you actually said/support/condone that.

- 做類似的評論前最好先謹慎評估。
- You have to be careful when you make statements like that.

- 你剛剛說的話很容易得罪人。
- A statement like that can land you in hot water.

- 你有聽見自己說的話嗎？
- Did you even hear what you just said?

- 我想我們應該檢討一下你的溝通方式。
- Maybe we should talk about the way in which you communicate.

- 你真的應該要注意自己在公開場合的發言。
- You should really watch what you say in public.

- 如果你真的這麼想，那我也沒什麼好說的。
- If you think like that, there's not much I can do to help you.

- 你總是想到什麼就脫口而出嗎？
- Do you always say everything that pops into your head?

- 我聽見了，也知道你想說什麼。
- I distinctly heard what you said and know that you meant it.

- 你知道自己講了不該講的話嗎？
- Do you realize you said that out loud?

- 說這些話，你很得意嗎？
- Do you take pride in this?

・你的發言很不恰當。

・This is not the right place to talk about this.

・你真的不應該說出這些話。

・There are things that just shouldn't be said.

・我沒有心情聽你說這些。

・I'm so not in the mood for this.

・你是不是存心想要激怒／挑釁別人？

・You like to upset/shock/unnerve people, don't you?

・如果這是你的看法，那還需要和你討論嗎？

・If that is your stance, what is there left to talk about?

・如果你對自己的想法很滿意，那我也沒意見。（諷刺語氣）

・If you are happy believing that, I guess I'm happy for you. [sarcasm]

・你的溝通技巧這麼好，實在應該去做外交官的。（諷刺語氣）

・You should be working as a diplomat with skills like those. [sarcasm]

・我沒辦法接受你這種發言／態度。

・I will not tolerate this kind of talk/treatment/attitude!

045 當對方生氣時

委婉	
・人生難免有不順的時候，不如先冷靜冷靜。	・We all have bad days—let's take a second to cool off.
・我知道你只是在氣頭上，不是真心的。	・I'm sure you didn't mean that.
・我知道你沒有要冒犯其他人的意思。	・I know you don't want to offend anybody.
・我們的態度應該盡量專業／有修養。	・Let's keep this professional/sophisticated.
・你說得有些過頭了。	・You're taking your point a little too far.
・現在不是情緒失控的時候。	・This isn't the time to let emotions run away from us.
・你有點反應過度了。	・You're getting a little intense here.
・我覺得你是故意要刺激我，是不是？	・I feel like you're trying to hurt me—is that the case?
・我感覺你的口氣有點挑釁／衝。	・I find your tone to be a bit provocative/upsetting/confrontational.
・請不要這樣對我說話！	・Please don't speak to me that way.
・沒必要涉及人身攻擊吧！	・There is no need to get personal.
・你正在氣頭上，該去冷靜一下。	・You've reached your boiling point; you need to settle down.
・我先把話說清楚，你最好不要這樣和我講話。	・I'm telling you up front: it's best to avoid that kind of talk.
・你已經說得太過分了。	・Your tone has crossed the line.

- 我想你應該去走一走／呼吸一下新鮮空氣／冷靜一下。
- Maybe you should take a walk/get a little fresh air/calm down.

- 如果你不能控制自己的脾氣，我們下次再說。
- If you can't control your temper, we'll need to talk some other time.

- 請閉上你的鳥嘴。
- Use your inside voice, please.

- 我不能接受你這種行為／言語／態度。
- I won't tolerate this kind of talk/behavior/tone.

- 你是故意這麼混帳的嗎？
- Are you trying to be mean?

- 你可以浪費你自己的時間來發脾氣，但請不要浪費我的時間。
- You may have a right to be angry on your own time, but not on mine.

- 你的字句／語氣根本令人無法接受。
- Your language/tone is offensive and unacceptable.

- 如果你再不停止，就請你滾出去。
- Keep a lid on it or you'll be out of here.

對立
- 你給我滾！
- Knock it off!

046 被緊迫盯人時

客氣

- 我非常樂意，但是我真的沒辦法／沒時間／要先離開。
  - I'd love to, but I really can't/have to go/don't have the time.

- 和你談話真是愉快，但我真的有事得先離開了。
  - It's been a pleasure, but I really have to go.

- 抱歉，但現在真的不是時候。
  - This simply isn't the right time, unfortunately.

- 我很想幫忙，但現在真的是分身乏術。
  - I would love to help you, but I'm already overcommitted as it is.

- 我希望能夠幫上更多忙，但目前真的沒有辦法。
  - I wish I could be of more help to you, but I can't.

- 我必須忍痛拒絕你的要求。
  - I must respectfully decline.

- 抱歉，但我真的幫不上忙。
  - I'm sorry, but I can't help you.

- 我現在沒時間處理這件事。
  - I don't have time for this at the moment.

- 請不要打擾我。
  - Please leave me alone/stop pushing.

- 不管你再怎麼說都是沒用的。
  - This conversation is unnecessary.

- 也許這是你的做事方式，但我不喜歡這樣。
  - Maybe this is the way you get things done, but I don't like it.

- 不用花力氣在我身上，沒有用的。
  - It's useless; you're wasting your efforts on me.

- 你這樣一直逼我，也只是浪費你的時間而已。
  - You'll be wasting your time if you keep talking to me/pushing me.

- 你為什麼不去找其他有時間的人？
  - Why don't you take this up with someone who actually has the time?

- 你真的不應該浪費精力在我身上。
- You really should put your energy elsewhere.

- 如果我需要你的幫忙，我會開口。
- If I need your help, I'll ask for it.

- 我的答案是，不。
- The answer is no.

- 我說不就是不。
- No is a complete sentence.

- 你別太過分，不要再煩我了。
- You're pushing too hard—stop it!

粗魯
- 拜託你去煩別人吧！
- Go bother somebody else!

047 當他人固執己見時

專業
- 這件事或許還可以從其他角度來看。
- There is probably more than one way to look at this.

- 還有別的辦法可以處理這件事。
- There are other ways to deal with this issue.

- 我了解你為什麼這麼認為，但是我真的無法同意。
- I can see where you're coming from, but I still must disagree.

- 如果你能客觀的看這件事，就會發現我是對的。
- If you thought about this objectively, you would find I'm right.

- 我完全了解你的想法，但我的看法和你不同。
- I certainly understand how you see it, but I see it differently.

- 我想你可能不是很了解我的想法。
- I'm afraid you didn't understand me very well.

- 你願意聽一聽我的看法嗎？
- Are you open to hearing what I have to say?

125

- 如果你始終堅持你對我錯，那我們根本沒辦法繼續討論下去。
- If you dig in and say, "I'm right and you're wrong," we won't get anywhere.

- 如果我們誰也不讓步，就只能一直僵持下去了。
- We'll just stay deadlocked forever if neither of us will concede.

- 如果我們雙方都堅持自己是對的，一定無法達成共識。
- When we both just want to be right, we won't accomplish anything.

- 請靜下心來理智地想想。
- Please listen to reason for a moment.

- 你得虛心接納別人的看法才行。
- You need to be open to others people's perspectives, too.

- 討論的目的就是要分享並且聆聽彼此的意見。
- The goal of any discussion is to share opinions and be open to the opinions of others.

- 如果我們兩個都只想贏，那只會兩敗俱傷。
- If both of us want to come out on top, neither one of us will.

- 如果永遠堅持自己是對的，任何討論都不會有結果。
- There's no "I'm always right" in a productive discussion.

- 如果你總是要爭到贏為止，那麼我們兩個根本不可能達成共識。
- We won't make any progress if you always want the last word.

- 希望你能明白，你不可能永遠是對的。
- You're not always right; I hope you can see that.

- 我們不要再繼續幼稚地爭吵下去。
- Let's stop acting like children.

- 為什麼所有事都要照你的意思啊？
- Why is it always your way or the highway?

- 如果你不能理性地討論，我也不想再說了。
- If you can't listen to reason, I'm done, here.

- 你別這麼固執，冷靜地聽別人說啊！
- You're just being a blockhead—listen to reason!

- 冥頑不靈對事情一點幫助也沒有。
- Being pig-headed never helped anyone.

- 既然你這麼堅持，想必你一定是對的。（諷刺語氣）
- You must be right since you are so sure. [sarcasm]

- 你這輩子有認錯的時候嗎？（諷刺語氣）
- Are you ever wrong? [sarcasm]

直接
- 你頑固地像頭牛。
- You're as stubborn as a mule.

048 當他人態度傲慢

客氣
- 不好意思，想請教一下你為什麼用這種態度和我說話？
- I'm sorry to ask you this, but why are you speaking to me like that?

- 我不太了解你說的話，請解釋一下你想要表達什麼？
- I'm not sure how to take your comment; would you please explain the intent?

- 你可以換一個方式再說一次嗎？我想我可能誤會了你的意思。
- So that I don't misinterpret your statement, would you please rephrase it?

- 你剛說的話很傷人，請問為什麼要這麼說？
- Your remarks are hurtful. Would you help me understand where they're coming from?

- 我是不是聽錯了？
- Am I hearing you correctly?

- 不曉得我是不是誤解了你的意思？
- I wonder if I'm misunderstanding what I'm hearing.

- 沒有人喜歡被這樣貶低。
- No one likes being talked down to.

- 你可能不了解自己說的話有多傷人。
- You may not realize how much your remarks hurt.

- 你剛剛是不是說〔對方說的內容〕？
- Did I just hear you say [repeat the phrase back]?

- 你是存心想羞辱我？還是有什麼其他目的？
- Are you trying to insult me or is there something else going on here?

- 你能不能對別人稍微友善一點啊？
- A little kindness would suit you better.

- 我本來想和你好好溝通的，但你似乎不這麼想。
- I wanted to talk things over with you, but it seems you're not open to that.

- 我希望你能夠更尊重我一點。
- I would appreciate a little more courtesy.

- 沒有人會認同你這種說法。
- Not everyone would agree with you, there.

- 言談之間貶低別人不是好的行為。
- It's not very nice to talk down to people.

- 我希望我們能找出合作的辦法，別再互相傷害了。
- I'd really like us to figure out a way to work together that doesn't hurt so much.

- 我本來以為你是個心胸開闊的人，沒想到你這麼瞧不起人。
- That's so condescending—I thought you were a bigger person than that.

- 你這種態度不會有好結果的。
- Nobody wins with that attitude.

- 請不要這樣和我說話。
- Please don't speak to me that way!

- 你為什麼總要把場面搞得這麼難看？
- Why do you always make our encounters so difficult?

- 你這樣對別人，一定會自食惡果的。
- Treating people like that will get you nowhere.

- 我們兩個是平等的，請注意你的態度！
- That's no way to act. My opinion is as valuable as yours!

- 如果你打算傷害我或侮辱我，我是不會受影響的。
- You're trying to intimidate me/put me down; well, it isn't going to work.

- 你對於不同意你的人都是這種態度嗎？
- Do you always give an attitude to those who don't agree with you?

- 你不用因為我們不同意，就態度這麼惡劣。
- You don't have to be so contemptible, just because we don't agree.

- 你為什麼老是對持不同意見的人這麼嗤之以鼻呢？
- Why do you always look down your nose at anyone with a different opinion?

- 你憑什麼覺得你可以這樣羞辱我？
- Who do you think it's okay to insult me like that?

- 你說話的態度可以不要這麼惡劣嗎？
- Can you talk without being quite so rude?

- 你有沒有教養啊？
- Don't you have any manners?

- 跟你說話簡直是浪費時間。
- There's really no use talking to you, is there?

- 我希望我也能夠像你這麼完美。（諷刺語氣）
- I wish I could be as perfect as you are. [sarcasm]

- 你別用那種居高臨下的態度對我。
- Don't cop that "holier than thou" attitude with me.

- 你以為你是誰啊？
- Just who do you think you are?

粗魯
- 難怪你一個朋友都沒有！
- No wonder you don't have any friends.

049 當他人出言威脅

技巧	
· 你為什麼對我有敵意？	· Why are you being so hostile?
· 我不喜歡你剛剛的發言。	· That's an unwelcome statement.
· 你已經有些過分了，我相信你私下會反省的。	· That's crossing a line—I'm sure you'll rethink things.
· 我們都是專業人士，不應該這麼衝動。	· There's no reason to act like that—we're professionals, you know.
· 你是在激怒我嗎？	· Are you trying to upset me?
· 你剛剛說的話不僅多餘，而且很傷人。	· Your words are unnecessary and hurtful.
· 請你停止。	· I would like this to stop.
· 這根本是浪費時間，我先離開了。	· This is a waste of time; I'm walking away.
· 你確定你要用這種態度和我說話嗎？	· Are you sure you want to speak to me this way?
· 別衝動說出自己會後悔的話。	· Don't say something you'll regret later.
· 口出威脅和肢體暴力一樣會造成傷害。	· Threats are just as harmful as actual violence.
· 你該聽聽自己說了什麼話！	· You don't know what you're saying.
· 不管怎樣，你不應該說出這種威脅的話。	· There is no excuse for incendiary comments like that.
· 我沒時間聽這些。	· I don't have time for this.
· 如果我是你，就不會來惹我。	· I wouldn't provoke me if I were you.

- 每個人都有該守的分寸，你剛剛已經越界了。
  - There's a line of no return in every relationship, and you've just crossed it.

- 你昏頭了嗎？居然會說出這些話！
  - Your wits have obviously left you.

- 你精神有問題吧！
  - You're obviously delusional.

- 你再說下去，我馬上打電話檢舉你。
  - I won't hesitate to call the authorities if you continue in this vein.

- 威脅我絕對是個不智之舉。
  - Threatening me is not something you want to do.

- 你知道要為剛剛那些話付出多大的代價嗎？
  - Are you aware of how much this is going to cost you?

- 你明白這麼說的後果嗎？
  - Do you know what you're risking?

- 我想你不知道你的話會造成了多大的傷害。
  - I don't think you realize the damage you've already caused.

- 你以為區區幾句話就能嚇唬我嗎？
  - You can't scare me with mere words.

- 如果你以為我會怕，那就大錯特錯了。
  - If you think I'm frightened, you're mistaken.

- 你真是風趣啊！（諷刺語氣）
  - You make me laugh. [sarcasm]

- 我受到威脅的時候，什麼手段都使得出來。
  - I won't hesitate to use force if I'm threatened.

- 必要時我一定反擊。
  - I will counterattack if necessary.

- 你有膽過來再說一次。
  - Come over here and say that again.

- 我現在就報警。
  - I'm calling the cops.

直接
- 你去吃屎吧！
  - Kiss my ass.

050 當有人對你百般挑剔

風度

- 嗯，我只能說我無法同意你的說法。（玩笑口吻）
- Well, I'll just have to disagree with you, there. [joking]

- 每個人都有不同的看法，如果剛剛說的有不對的地方，請見諒。
- We all have different opinions; I'm sorry if I misspoke.

- 我只是想提出具體的建議，請別誤會。
- I meant that to be constructive. Please don't take it any other way.

- 我剛剛說的只是就事論事／一般好意。
- My words were meant in a professional/ kindly manner.

- 或許我應該說得更得體一些，但我相信你了解我的意思。
- There are better words than the ones I chose; however, I know you understood me.

- 也許我表達得不夠好，但我想你應該了解我的意思。
- Maybe my words were not the best, but you understand what I am trying to convey.

- 抱歉，我剛剛的用字遣詞不夠精確。
- Sorry, I didn't think using such precise language was critical just now.

- 我們都是受過教育的人，請不要對我剛剛的話耿耿於懷。
- We're civilized people; what I've said shouldn't make us enemies.

- 我只不過是陳述自己的想法。
- I was merely stating my opinion.

- 我只是說…。
- I was just saying....

- 我沒想到你會這麼介意。
- I didn't expect you to take offense.

- 我並沒有要說服任何人的意思。
- I'm not here to convince anybody.

- 我只是把自己的觀察說出來。
- I was just making an observation.

- 我的意見和你一樣有分量。
- My opinion is just as valid as your opinion.

- 我只是做一個簡單的陳述。
- I was just making a simple statement.

- 我可以把話講得很好聽，但我認為沒有這個必要。
- I could sugar-coat it, but I don't think that's necessary.

- 你把事情想得太嚴重了。
- I think you're taking this too seriously.

- 你沒必要為此發這麼大的脾氣吧！
- There's no need to work yourself up into a state.

- 你的臉皮未免太薄了些。
- You shouldn't have such thin skin.

- 挑我毛病對事情並不會有任何幫助。
- Picking on me isn't going to change anything.

- 我說的或許不是百分之百正確，但我的論點有一定的道理。
- My words may be imprecise, but I know what I said made sense.

- 我可以換句話說，但你已經了解我的意思，還有這個必要嗎？
- I could say it differently, but why would I? You get what I'm saying.

- 如果你用心理解我剛剛說的內容，就不會有這麼多的不滿了。
- You wouldn't have any reason to pick on me if you made an effort to understand me.

- 你想這樣的話，我可以花上一整天對你解釋我剛剛說的話。
- I'm quite happy to spend all day defending my words, so you'd better get comfortable.

- 我只是說出我的想法而已，有錯嗎？
- I just spoke my mind—what's the problem?

- 為什麼你不能接受我們是兩個想法不同的個體呢？
- Why can't you accept that we are two people who think differently?

- 接受我們兩個不同的意見有這麼難嗎？
- Why is it so hard to accept that we have differences?

· 不要這麼敏感好嗎？生活中還有很多其他事情。	· Stop being so sensitive; there's a big world out there.
· 我一向有話直說，如果你不欣賞，我也愛莫能助。	· I tend to say what's on my mind—if you don't like it, I can't help you.
· 和你說話之前是不是應該先讀一下說明書啊？（諷刺語氣）	· Are there instructions on how one ought to communicate with you? [sarcasm]
· 我很遺憾你的智商無法理解我說的話。	· I'm sorry if you were just too stupid to understand my meaning.
· 請接受我誠摯的道歉，國王陛下。（諷刺語氣）	· My apologies, Your Majesty! [sarcasm]

粗魯

---

051 當有人說謊

友善

· 你是存心扯我後腿，對吧？（玩笑口吻）	· You're pulling my leg, right? [joking]
· 看來有人是放羊的小孩喔！（玩笑口吻）	· Someone's pants are on fire! [joking]
· 誠實才是上上策。（責備或玩笑口吻）	· Honesty is the best policy. [scolding/joking]
· 你不覺得，誠實是最好的策略嗎？	· Honesty is always the best policy, don't you think?
· 我真希望我能相信你說的話。	· I really wish I could believe what you're saying.
· 我尊重說話誠實的人。	· People who are straight with me garner my respect.

・我知道這不容易，但請對我說實話。 ・I know it's hard, but please be straight with me.

・你不說實話只會讓事情變得更糟糕。 ・That's only making a delicate situation worse.

・說出真相你才能夠解脫。 ・The truth will set you free.

・說謊是很不好的行為。 ・It's not fair to mislead people

・你我都知道你說的不是真的。 ・You and I both know that's not true.

・你何不乾脆告訴我實話？ ・How about telling me the truth instead?

・你說的內容並不合理。 ・Something doesn't add up here.

・這件事聽起來就有鬼。（玩笑口吻） ・Something's rotten in the state of Denmark. [joking]

・我怎麼知道你在說謊啊？看你動嘴皮的樣子就知道啦。（玩笑口吻） ・How do I know you're lying? Because your lips are moving. [joking]

・說一個謊，要用無數的謊來圓前面的謊。 ・What a tangled web we weave, when first we practice to deceive.

・說謊終究是會被拆穿的。 ・Lying won't get you anywhere.

・這樣只是讓事情越來越糟。 ・You're only making things worse.

・我討厭謊言，更討厭說謊的人。 ・I hate lies—and liars.

・看著我的眼睛再說一次。 ・Look me in the eyes and say that again.

・你不如試著誠實一次。（諷刺語氣） ・How about being honest for a change? [sarcasm]

挑釁

052 當有人想和你吵架

化解

- 我真的不想把事情鬧大，如果我說了什麼得罪你的話，請多見諒。
- I really don't want this to escalate—I'm sorry if I said the wrong thing.

- 如果我的行為冒犯了你，我很抱歉。
- If this is going to cause an altercation, I take it all back.

- 我真的沒有任何冒犯你的意思。
- I certainly didn't mean to offend you.

- 我希望剛剛說的話沒有造成任何衝突。
- I hope what I said won't cause a rift between us.

- 我希望你能大人不記小人過。
- I'd prefer to take the high road, here.

- 我知道你不是真的想吵架。
- I know you don't really want to pick a fight.

- 讓我們就事論事／保持專業態度。
- Let's keep things professional/sophisticated/on the up and up.

- 不管任何關係都需要彼此尊重。
- Any relationship we have must be based on respect.

- 真正專業的人會先把自己私人的情緒放在一邊。
- A true professional puts his/her personal feelings on the back burner.

- 我們還有時間挽回場面，就看你願不願意。
- There's still time to salvage the situation; it's up to you.

- 你真的不需要有這麼大的反應，用點理智吧！
- You don't have to act this way; a reasonable person wouldn't.

- 在職場上，請表現的像個專業人士。
- You're a professional—act like one.

- 我不打算和你一般見識。
- I will not lower myself to your level.

- 你可以小聲點／口氣好一點／更尊重別人一點嗎？
- Would you please lower your voice/change your tone/speak more respectfully?

- 不要只想到自己，有時也要顧慮到他人。
- Think of someone other than yourself!

- 別鬧了，我是不會陪你瞎攪和的。
- Knock it off—I'm not going to engage with you!

- 如果你再這樣，我們只好請你離開。
- Keep this up and you'll be on the outside looking in.

- 如果你繼續這麼幼稚，今天就到此為止。
- Let's stop now if you're going to be immature about it.

- 你再繼續這種行為，就請你另外找工作／朋友／伴侶。
- Keep acting like this and you'll be looking for another [job/friend/spouse].

- 你就是想找架吵嗎？
- Are you looking for a fight?

- 我不會允許任何人這麼對我說話。
- I don't allow anyone to treat me this way.

- 如果你說不出什麼好話，就請閉嘴。
- If you can't say anything nice, keep your mouth shut.

- 看來有人今天打算自找苦吃。
- Looks like someone is cruising for a bruising.

053 當被人打斷時

溫和

- 請讓我把最後這部分做完。
  - Please allow me to add just one more thing.

- 抱歉，我還差最後一個步驟就要完成了。
  - Sorry, I've got one last point to make.

- 請讓我處理完最後一小部分，然後就是你的時間了。
  - Please just let me finish and then you'll have the floor, I promise.

- 請讓我繼續處理手邊的事。
  - Please let me continue.

- 我希望能繼續處理原本的事。
  - If I could just go on.

- 你讓我無法完成工作。
  - You're not letting me finish.

- 我可以先專心在本來的事情上嗎？
  - May I finish my thought?

- 你可不可以先讓我把事情處理一個段落？
  - Won't you allow me to conclude my point?

- 你可以讓我把話講完嗎？
  - If you would just let me get a word in edgewise.

- 我還沒完成。
  - I wasn't done yet.

- 如果你能先聽我說完，很多疑問都能獲得解答。
  - If you'd just listen, all your questions would likely be answered.

- 和你溝通很困難，因為你會不斷插話。
  - I have a very hard time communicating with you when you keep cutting me off.

- 如果你不介意，請先讓我把話說完。
  - I'd like to finish speaking if that's okay.

- 你似乎不想聽我說的話。
  - You don't seem to want to hear what I have to say.

- 我可以聽你說，但可以先讓我講完要說的話嗎？
- I'm okay listening to you, but are you okay letting me finish?

- 先聽我講三十秒，然後我專心聽你說。
- I let you speak; now please listen to me for 30 seconds.

- 你根本不在乎我要講什麼，對吧？
- What I say doesn't really matter to you, does it?

- 你是不是完全不想聽我說話啊？
- Do you care at all about what I have to say?

- 你如果要聽我說，就不要一直打斷我。
- For you to hear me, you'd first have to let me speak.

- 你為什麼要一直打斷我說話呢？難道怕我說出什麼嗎？
- Why do you interrupt all the time? Are you afraid of what I have to say?

- 像你這麼會說話，應該也要會聽別人說話才對。
- For someone who talks so much, you should know how to listen.

- 輪到你的時候，自然會讓你講。
- When it's your turn, I'll let you speak.

- 請先等我講完。
- Hold your peace until I'm done.

- 不要在我還沒講完前插話。
- Don't talk until I'm finished.

- 拜託讓我說完。
- Please let me speak!

- 安靜，聽我說！
- Be quiet and listen to me!

直接
- 拜託你閉嘴。
- Shut up already!

054 當被人取笑時

客氣

- 我很高興能帶給你這麼大的歡樂，現在我們可以繼續討論嗎？（玩笑口吻）
- I'm glad you had a laugh at my expense—can we move on? [joking]

- 拜託，你就承認忌妒我吧！（玩笑口吻）
- Come on; just admit that you're jealous! [joking]

- 如果你不知道怎麼開玩笑，我可以教你。（玩笑口吻）
- I can teach you a thing or two about jokes. [joking]

- 什麼事這麼好笑？
- Hey, what's so funny?

- 你知不知道你的笑話很傷人啊？
- Did you know that your humor can be hurtful to some people?

- 取笑我只會讓你自己難堪而已。
- You're just embarrassing yourself, you know.

- 你是不是很喜歡取笑別人啊？
- You like laughing at people, don't you?

- 我不欣賞這種低級幽默。
- I don't appreciate the low blow.

- 這裡不是你可以胡說八道的地方。
- It's no place for that kind of nonsense.

- 我不喜歡你這種行為。
- I don't like how you're behaving.

- 顯然，你的幽默感有問題。
- Obviously, humor doesn't come naturally for you.

- 這是你表示幽默的方式嗎？
- Trying to be funny again?

- 你覺得自己有資格說這些話嗎？
- What makes you think you can say that?

- 我不喜歡你對我講話的方式。
- I don't like the way you're treating me.

- 為什麼你老是喜歡貶低別人呢？
- Why do you insist on bringing other people down?

- 也許你應該把這套高標準拿來要求自己。
- Maybe you should put those expectations of perfection on yourself.

- 這樣取笑我，你心情很好嗎？
- Do you feel better now that you've laughed at me?

- 你的愚蠢真的是高人一等。
- You've got a real talent—for stupidity, that is.

- 你這種行為只會讓自己看起來很幼稚。
- You're being immature, even for yourself.

- 難看的人不是我，而是你。
- You're the one who looks ridiculous here, not me.

- 真是有創意。（諷刺口吻）
- How original. [sarcasm]

- 我一直都很欣賞你高明的談話技巧。（諷刺口吻）
- I've always appreciated the way you make people feel comfortable. [sarcasm]

- 很抱歉我不像你這麼聰明。（諷刺口吻）
- I'm sorry I'm not as clever as you. [sarcasm]

- 真高興我能娛樂到你。（諷刺口吻）
- So glad I could entertain you. [sarcasm]

- 好奇怪，同樣一句話，其他人說出來就幽默多了。（諷刺口吻）
- I've heard that before, but from someone with an actual gift for humor. [zinger/sarcasm]

- 有時候我真懷疑你到底有沒有大腦。（認真口吻）
- Sometimes I wonder if you have a brain at all. [musing]

- 好棒，你好專業啊！（諷刺語氣）
- Bravo, how professional. [sarcasm]

- 我不回應，因為我不想佔沒智商的人的便宜。
- It's not fair to have a battle of wits with an unarmed person—so I won't reply to that.

粗魯

055 當被人批評時

接納

- 你應該也是掙扎了很久才說的吧？謝謝你這麼坦白。
  - That must have been hard to say; I appreciate your honesty.

- 你的意見很重要，謝謝你告訴我。
  - Your opinion means a lot to me; thank you for offering it.

- 我很感激你的回饋，會記在心裡。
  - I appreciate the feedback and will take it to heart.

- 你的話深深啟發了我，真是謝謝。
  - You gave me a lot to think about, I appreciate that.

- 非常感謝你告訴我這些。
  - Thank you so much for letting me know about this.

- 我很高興你特地告訴我這些話。
  - I'm glad you took the time to tell me these things.

- 我一定會認真參考你的意見的。
  - I will certainly take what you said into consideration.

- 我一直都很樂意接受大家的建言。
  - I always like to hear a lot of opinions.

- 你似乎沒辦法欣賞和你不一樣的人，我們可以討論一下嗎？
  - You seem really uncomfortable with differences—can we talk about that?

- 我喜歡有建設性的意見，但你說的似乎有點過火。
  - I like constructive criticism, but this seems a bit harsh to me.

- 通常你的意見都很中肯，但這次似乎是例外。
  - Your comments are usually right on, but they're not justified in this case.

- 我不認同你說的話。
  - I don't think that's fair to say.

- 你可以接受和你不同的行事風格嗎？
  - Is it okay if people do things differently than you?

- 這些話似乎不應該在這個場合裡出現。
- This isn't the right place to call someone out like that.

- 每個人都有自己的缺點。
- To each person his or her own flaws.

- 通常批評別人的人都是五十步笑百步。
- When you're pointing a finger at someone, you have four fingers pointing back at you.

- 我沒有辦法讓每一個人都滿意。
- You can't please everyone all of the time.

- 你不覺得自己有點誇大其詞了嗎？
- Don't you think you're exaggerating, here?

- 你總是批評不合自己心意的人嗎？
- Do you usually criticize people who are different than you?

- 批評別人之前應該先檢討自己吧！
- You should take a look at yourself before you start judging others.

- 我不知道你為什麼說出這麼刻薄的話，請不要這樣和我說話。
- I don't know where all this acrimony is coming from, but please don't direct it at me.

- 你這麼做只是讓自己難看而已。
- Doing this only brings you down in the eyes of others.

- 你先把自己做好再來說別人吧！
- People who live in glass houses shouldn't throw stones.

- 這件事情上，我不接受他人的批評。
- I won't allow any judgments on that.

- 恕我不回應。
- I would never allow myself to comment on that.

- 我不打算配合你演出。
- I won't dignify that with a response.

- 你有真心接納過別人嗎？
- Don't you ever accept others as they are?

- 如果你不歡迎我，就直說。
- If I'm not welcome here, just let me know.

- 如果你不想看到我，隨時歡迎你離開。
- You don't have to stick around if I bother you so much.

- 如果你對我不滿，乾脆挑明說清楚。
- If you have a problem with me, let's just get it out in the open.

- 如果你的眼裡容不下一粒沙子，那我們恐怕很難相處。
- If you're looking for perfection, you shouldn't hang out with me.

- 你憑什麼批評我？
- What gives you the right to judge me?

- 你覺得自己真的有資格這麼說嗎？
- Who dictates the code of behavior here? You?

- 在批評我之前拜託先照照鏡子吧！
- Before you criticize me, take a look in the mirror.

- 你真的覺得自己是個好榜樣嗎？
- Do you think you're a great example?

- 我只接受我尊敬的人的評語。
- I would acknowledge your comments if I respected you.

- 這裡唯一出問題的就是你。
- There's only one problem here and it's you!

- 繼續罵啊，讓全世界知道你有多完美。（諷刺語氣）
- Just keep on criticizing me; the entire world knows how perfect you are. [sarcasm]

回擊

- 如果要像你一樣才能讓你認同，那我不要！
- If I have to be like you to gain your respect, I don't want it.

056

當他人防衛心過高

安撫

- 如果我冒犯了你還請見諒，我不是有意的。
  - Please forgive me if I stepped on your toes; I did not mean to upset you.

- 假如說話傷到了你，我很抱歉，真的是無心之過。
  - I apologize if I hurt your feelings; it was not my intention.

- 我沒有要冒犯你的意思，只是想開個玩笑而已。
  - I didn't mean to offend you; it was only meant in jest.

- 拜託，不要想得這麼嚴重嘛！
  - Come on, please don't take it like that.

- 我會這麼說也是為你好／也是關心你。
  - I was saying it for your benefit/because I care about you.

- 為什麼我覺得自己好像惹你生氣了？
  - Why do I get the feeling that I'm disturbing/upsetting you?

- 為什麼防衛心這麼重呢？
  - Why the defensiveness?

- 你為什麼把事情想得這麼嚴重？其實真的還好。
  - Why are you taking this so hard? It's not as bad as you think.

- 不管是哪一種人際關係，學習接納他人意見都很重要。
  - Giving and accepting advice is part of any relationship.

- 我們等你心情好點／冷靜一點再來討論這件事。
  - Let's discuss this when you feel better/calm down/can see more clearly.

- 我不懂你為什麼要這麼激動啊？
  - I don't understand why you're expressing yourself this way.

- 這件事真的沒什麼，沒有必要搞這麼僵吧！
  - There's no need to be so defensive—it's not that big a deal.

- 你為什麼要這樣解讀這件事情呢？
  - Why are you taking it like that?

- 我明明不是這個意思，你為什麼一定要認為我有呢？
  - Why do you attribute motives to me that don't exist?

145

- 你的臉皮太薄了。
- You shouldn't be so thin-skinned.

- 我發現你從來不接受別人的批評。
- I see that you never let others disagree with you.

- 我發現你從來不聽和你不同的聲音。
- I see that you never let anyone contradict you/speak into your life.

- 你為什麼這麼暴躁？
- Why are you so irritable?

- 你何必在心裡築起這麼多道牆呢？
- Why do you put up so many walls?

- 你為什麼總是對我這麼防備啊？
- Why must you always stonewall me?

- 你根本是小題大作嘛！
- You're making a mountain out of a molehill.

- 如果你對自己更有信心，就不會對這件事有這麼大的反應了。
- If you were more secure with yourself, you wouldn't be so bummed about this.

- 這完全不是針對你，如果你要那樣想，我也無可奈何。
- This isn't an attack on you—it's a general statement, so too bad if you took it personally.

- 你真的不要再鑽牛角尖了。
- You really need to let this go.

- 並不是每個人都像你這麼完美。（諷刺語氣）
- If only we were all perfect like you. [sarcasm]

- 我了解，實話總是很傷人的。
- The truth hurts, I'm sure.

- 你反應這麼激動，我完全沒辦法和你溝通。
- I can't reason with you when you're like this.

- 有時候你真的是無理取鬧。
- It's impossible to have a conversation with you sometimes.

防衛

- 如果你需要心理諮商，我可以推薦。
- There's an 800 number I can recommend if you feel you need help with this.

057

當被人質疑時

正面

- 你知道我永遠不會辜負你的信任。
  - You know you can always count on me.

- 我是你在這裡唯一可以信任的人。
  - If there's anyone you can trust here it's me.

- 我是你最不應該懷疑的人。
  - I'm the last person you should be doubting.

- 這是我第一次被人懷疑。
  - It's my first time in a situation like this.

- 你不是在懷疑我吧？
  - You aren't looking at me, are you?

- 我以前遇過類似的情況，你有嗎？
  - I've been through this before—have you?

- 那麼你希望我怎麼做？
  - What do you expect from me?

- 很遺憾之前的錯誤影響了你對我的觀感。
  - Sorry if my past mistakes affected your opinion of me.

- 你要懷疑就懷疑吧！
  - I give you the benefit of the doubt.

- 我自己也不是很有把握。
  - I've got that sinking feeling, too.

- 我也不是非常確定。
  - I'm sensing the uncertainty, too.

- 聽著，雖然現在看起來很糟糕，但是會越來越好的。
  - Listen, it may seem bad now, but it will get better.

- 我自己也不放心。
  - I'm not comfortable with it, either.

- 你到底有什麼意見？我的計畫是無可挑剔的。
  - What's the problem? My plan is unimpeachable.

- 如果你要找人懷疑，去照照鏡子吧！
  - If you want someone to doubt, look in the mirror.

防備

- 或許你自己才是有問題的人。
  - Maybe you're the one with the problem.

147

058 當有人逃避你的問題

客氣

- 所以答案是…？（玩笑口吻）
- And the answer is…? [joking]

- 如果閃躲問題是奧運項目的話，你一定是個金牌得主。（玩笑口吻）
- If avoidance were an Olympic event, you'd win the gold medal. [joking]

- 看來這個問題很難回答，我換個方式問你好了。
- It seems like we're having trouble focusing—let me say it another way.

- 好吧！我換個方法問你。
- Okay, I'll put it a different way.

- 我們不要再顧左右而言他了。
- Let's not dance around the issue any longer.

- 請不要模糊焦點。
- Let's get to the heart of the matter.

- 請你盡量回答這個問題。
- Please try your best to answer me.

- 我希望你能給一個更明確的答案。
- I was hoping your answer might be a little more specific.

- 沒關係，我再問一次。
- Okay, I'll say it again.

- 請你不要迴避我的問題。
- Stop evading me, please.

- 請不要閃躲問題。
- Please don't dodge the question.

- 你沒有回答到我的問題。
- You didn't answer my question.

- 需要我再說一次問題嗎？
- Do I need to repeat myself?

- 你沒聽到我的問題嗎？
- Didn't you hear my question?

- 你不打算給我一個答案嗎？
- Aren't you going to give me an answer?

- 需不需要我再講一次剛剛的問題？
- Do you need to hear the question again?

- 針對這個簡單的問題，我希望你能給出明確的答案。
- I would like a direct answer to my simple question.

- 這是我最後一次重複我的問題了。
- I'll repeat the question, but only one more time.

- 你是不是不知道答案？
- Do you just not know the answer?

- 你根本是在逃避我的問題，對吧？
- You're dodging the question, aren't you?

- 我到底要問你幾次才夠啊？
- How many times do I have to ask you?

- 沒想到我問的問題這麼難回答。（諷刺口吻）
- I didn't realize that I was asking the impossible. [sarcasm]

- 你很會迴避我的問題喔！
- Nice way to avoid my question.

- 逃避並不能解決任何問題。
- Denial is not just a river in Egypt.

無禮
- 給我答案吧！
- I demand an answer!

059 當有人對你吼叫

圓滑

- 冷靜的溝通比較有效。
- · Speaking calmly will make communicating with me much easier.

- 或許我們應該先冷靜一下。
- · Maybe we need to take a breather, here.

- 你不需要抬高音量。
- · There is no need to raise your voice.

- 你不必大吼大叫。
- · Yelling isn't necessary.

- 我沒辦法和吼叫的人溝通。
- · It's very hard for me to communicate with people shouting.

- 你可以體諒別人一些嗎？
- · You could be a little more understanding.

- 試著多點溫柔吧！
- · Try a little tenderness.

- 我到底做了什麼，讓你這樣和我說話？
- · What have I done that you would feel the need to speak to me that way?

- 多學點溝通技巧，對你很有幫助的。
- · More diplomacy and tact would suit you better.

- 為什麼不能更有禮貌一點？
- · How about being courteous?

- 你應該表現出更好的修養。
- · How about being a little more civil?

- 你似乎火氣很大。
- · You seem to be getting a little hot under the collar.

- 幹嘛這麼大聲？
- · Why the raised voice?

- 我想你應該找個地方先冷靜一下。
- · I think you should go somewhere to calm down.

· 你要先冷靜一下嗎？	· Do you need to take a time-out?
· 你已經失去分寸了。	· You're going too far.
· 吼叫也沒有用。	· Yelling will get you nowhere.
· 先安靜一分鐘。	· Breathe for two seconds.
· 你為什麼對我有這麼大的敵意？	· Why be so hostile?
· 你何必讓自己在大庭廣眾下出醜呢？	· Why are you making such an scene?
· 其實，柔軟一些對你只會有好處的。	· You know, it doesn't hurt to be nice.
· 你好好的講，我就可以聽懂了。	· I also understand things when they are said softly.
· 你不要太過分了。	· You're crossing a line, here.
· 說真的，你需要冷靜點。	· Seriously, you should calm down.
· 請你有點格調，好嗎？	· Show a little class.
· 為什麼要這麼引人側目？	· Why make such a scene?
· 你知道自己在吼叫嗎？	· Are you aware that you're screaming?
· 大聲說話不表示你比較聰明，也不表示別人就必須同意你。	· Talking that loudly doesn't make you any smarter or more convincing.
· 這樣吼叫有用嗎？	· What is this supposed to accomplish?
· 你吼夠了沒？	· Is that it? Are you done yelling now?
· 我很驚訝你的溝通技巧這麼高明。（諷刺口吻）	· I'm surprised by even that much tact out of you. [sarcasm]

- 真是好口才。（諷刺口吻）
- What eloquence. [sarcasm]

- 你再這麼大聲的嚷嚷，我要走了喔！
- Keep talking that loudly and you'll be talking to an empty room.

- 我拒絕在這種情況下跟你說話。
- I refuse to let anyone speak to me like that.

- 我受夠了，再見！
- I've heard enough—bye!

直接 · 我沒有聾
- I'm not deaf!

060 當有人對你罵髒話

含蓄
- 哇，請注意一下你的用詞。（玩笑口吻）
- Whoa, easy with the language! [joking]

- 嘿！我的耳朵可是很清純的。（玩笑口吻）
- My virgin ears! [joking]

- 你平常不是這樣的。
- You're better than that.

- 多一點溝通技巧，會對你大有幫助。
- A little diplomacy can go a long way.

- 你可以換個方式表達吧！
- There are other ways to say that.

- 這種行為不適合你。
- This kind of behavior doesn't suit you well.

- 你這樣說很不好。
- It's not really nice to say that.

- 你沒必要這樣說話吧！
- That kind of talk is unnecessary.

- 我對你很失望。
- You've disappointed me.

- 你讓我很不舒服。
  - I'm not feeling comfortable with you right now.

- 我們可以等你冷靜一點再繼續談。
  - We can come back to this when you calm down.

- 失控並不能解決問題。
  - Losing control never solves anything.

- 你有必要這樣說話嗎？
  - Is it really necessary to say that?

- 對人有點禮貌吧？
  - How about being nice?

- 請有修養一點，好嗎？
  - How about being a little civil?

- 哇，你好沒禮貌！
  - Well, that's offensive!

- 拜託，請注意一下你的用詞遣字。
  - Watch your language, please.

- 或許你沒注意到，但是這裡每個人都很有禮貌。
  - In case you were not aware, we speak politely around here.

- 如果你講話非得這麼難聽，那請你閉嘴。
  - If you can't say something nice, don't say anything at all.

- 說這種話真是浪費你受過的教育。
  - You're belying your education/upbringing when you talk like that.

- 我不知道你這麼粗魯。
  - I don't understand how you can take pleasure in being rude.

- 每天怒氣沖天的，你一定很累吧？
  - It must be exhausting to carry around so much anger inside.

- 為什麼你的表現這麼令人討厭啊？
  - Why are you being so offensive?

- 你不能使用一般的詞彙說話嗎？
  - Can't you speak without being vulgar?

- 你說話有用大腦嗎？
  - Do you ever think before speaking?

- 講話這麼難聽，對你有什麼好處？
- What does it gain you to be nasty like that?

- 再這樣下去，沒有人會想和你說話的。
- Keep speaking like that and you'll be talking to yourself.

- 你的說話技巧真是讓我甘拜下風。（諷刺口吻）
- I'm overwhelmed by your abundant tact. [sarcasm]

- 恭喜你，你的智商可真是高啊！（諷刺口吻）
- Congratulations, how intelligent. [sarcasm]

激烈
- 給我閉嘴！
- Knock it off!

061 當有人想轉移話題

客氣
- 你說得很好，但我們晚點再回來討論這一點。
- Good point—we'll get back to that in a second.

- 讓我們先回到原本的主題上吧！
- Let's get back to what we were talking about.

- 我們先把眼前的事情討論完好嗎？
- Let's get back to the subject at hand.

- 我覺得我們有點離題了。
- I feel that we're slightly off topic now

- 在進入下一個主題之前，我們應該先把本來預定的題目討論完吧！
- Let's stick to the agenda before moving on to new topics.

- 我們應該還在討論〔本來的題目〕吧？
- I think we were still talking about [topic at hand], no?

- 這件事沒討論完就換另一件事，一點意義也沒有。
- There's no use moving on until we're finished with the subject at hand.

- 我想我們已經完全離題了。
  - I think we're veering too far afield from the issue at hand.

- 我們本來在討論的應該不是這件事吧！
  - It seems to me that we weren't speaking about that.

- 你好像想岔開話題喔！
  - You seem to be deflecting the main issue.

- 你似乎想趁亂帶過這個話題。
  - It seems like you're trying to muddy the waters.

- 這件事真的和今天的主題有關嗎？
  - Is this topic really along the same lines?

- 這件事和今天的討論毫無關聯吧！
  - That has nothing to do with the current conversation.

- 你好像在自說自話。
  - You're approaching this conversation as though it were a monologue.

- 剛剛的事情還沒講完，對吧？
  - I don't think we've resolved the issue yet, do you?

- 跟你講話簡直像在兜圈子。
  - Talking to you is like being in a revolving door.

- 我想你只是在顧左右而言他。
  - I think you're just trying to confuse the issue.

- 拜託別轉移話題。
  - Please don't change the subject.

- 不要岔開話題。
  - Don't try to divert the conversation.

- 我好像在跟一道牆講話。
  - I feel like I'm talking to a brick wall.

粗魯
- 我還沒說完呢！
  - I wasn't done yet!

062 當被人侮辱時

含蓄

- 你好像有些焦慮／過度勞累／憤怒，有什麼我可以幫忙的嗎？
  - You seem a little anxious/overworked/upset—what can I do to help?

- 哇，我們應該成熟一點吧！（玩笑口吻）
  - Whoa, let's keep this sophisticated! [joking]

- 我比較喜歡有禮貌的你。
  - I like the polite version of you much better!

- 哇，你這樣說話很不對。
  - Aw, that wasn't very nice.

- 我想你有點小題大做了，你說呢？
  - I think you're exaggerating a bit, here—don't you?

- 我真的不想和你吵。
  - I really don't want to argue with you.

- 拜託你注意自己的詞彙。
  - Please be more careful with your choice of words.

- 這個場合並不適合這樣指名道姓。
  - This isn't the right place to call someone out like that.

- 我到底做了什麼？讓你這樣對我。
  - What did I ever do to you to warrant that?

- 為什麼你這麼沒禮貌？
  - Why are you being so rude?

- 你有必要這麼說嗎？
  - Is it really necessary to say that?

- 你這樣說，我很難回應。
  - I can't do much when someone is speaking to me like that.

- 我一向很尊重你。到底是哪裡出了問題？
  - I always treated you with respect. What's the problem?

- 有些話不應該說。
  - There are some things that just shouldn't be uttered.

・我從來不允許別人這樣和我說話。	・I never allow anybody to talk to me this way.
・這已經太超過了。	・This is going too far.
・我沒必要忍受這些。	・I don't have to stick around for this.
・我再也不要聽到你這樣說話。	・I don't want to hear you talk like this ever again.
・你再不收斂，我就…。	・Keep on crossing the line and I'll [...]!
・如果你累了，去休息吧！	・If you're tired, go to take a nap.
・有話好好說，好嗎？	・Just tone it down, okay?
・給我閉嘴！	・Knock it off!
・好棒的溝通技巧啊！（諷刺口吻）	・Bravo, what diplomacy. [sarcasm]
・盡量侮辱我啊！這好像是我們今天的目的。（諷刺口吻）	・Have fun insulting me; obviously we're all here for that. [sarcasm]

直接

063 當有人對你暴力相向

安撫	· 暴力並不能解決問題。	· Violence is never the answer.
	· 我以為你會尊重我。	· I thought you had more respect for me.
	· 這種行為舉止對大家都沒好處。	· This behavior will not benefit anybody.
	· 我討厭你動手。	· I don't appreciate you hitting me.
	· 請不要再動手了。	· Please don't put your hands on me.
	· 拜託，不要碰我！	· Keep your hands to yourself, please.
	· 你一點也不尊重他人。	· You've definitely crossed the line of respect.
	· 我想你需要專業協助。	· I think you really need help.
	· 你顯然有嚴重的問題。	· Clearly you have a problem.
	· 你需要放鬆一點。	· You need to relax.
	· 你這麼做絕對會丟工作。	· You can lose your job for this.
	· 不准你再出手打人。	· This is the last time you'll ever do that.
	· 你的行為已經犯法了。	· What you did can put you in jail.
	· 現在就住手。	· Stop this right now.
	· 我不會再原諒你的行為。	· I will not tolerate that again.
	· 嘿，你太過分了。	· Hey, you've crossed the line!
	· 給我自制一點。	· Get a hold of yourself.
	· 你應該要接受治療／坐牢。	· You need to be in a mental institution/behind bars.

- 我打算和你斷絕關係。
  - I'm tempted to knock you into yesterday.

- 現在你開心了嗎？（諷刺口吻）
  - Are you happy now? [sarcasm]

- 你對每個人都是這樣嗎？
  - Do you treat everybody like this?

- 如果你再對我動手一次，我發誓會給你好看。
  - If you come at me again, I will retaliate, I promise.

- 你敢再碰我一次，我就叫警察／人事部門。
  - Touch me again and I'm calling the police/authorities/HR.

- 這可能是你這一輩子最大的錯誤。
  - That was quite possibly the worst mistake you've evermade in your life.

- 敢再一次，我會讓你生不如死的。
  - Do that again and I'll ruin your life.

- 我現在報警。
  - I'm calling the cops.

- 徹底的滾出這裡！
  - You're finished here.

威脅
- 你想領教我的空手道嗎？
  - Care to see my black belt?

處理衝突與憤怒

3

Part 4

展現外交手腕

在不卑不亢中展現出領袖風範，贏得對方信任，並且傳遞自信、尊重、權威，以及誠信，甚至運用模稜兩可的藝術達到想要的效果。

溝通哪些事

- 如何開啟一場討論
- 如何結束一場討論
- 如何開始一場演說
- 如何結束一場公開演說
- 如何指控他人
- 如何凝聚共識
- 如何處理意見不同的場面
- 如何避免敏感話題
- 如何做建議
- 如何回覆不想採納的建議
- 當有人為了某件事指責你
- 當你是謠言主角

>
>
> **外交就是讓他人接受你的想法。**
> ——丹尼爾‧法爾（DanielVare，三〇年代的義大利商人）

　　外交官必須具備特殊技能，才能夠負責艱鉅的國際溝通任務，而外交官同時需要參加許多公開活動，幾乎不太可能「低調行事」，如果你的工作也需要大量的外交技巧，不妨參考下面幾個小建議：

① 贏得對方信任

　　想像你是一名代表台灣政府的領事或大使。你的任務就是在一堆複雜的會議和活動中與其他代表討論，並且解決非常敏感但重要的議題，同時還得盡可能地維持與他國的友好關係。如果你能展現領袖風範，傳遞出自信、尊重、權威，還有最重要的誠信，其他人將會基於信任而提供重要的資訊。

② 練習溝通

　　溝通技巧就像肌肉一樣需要訓練。你可以上演說訓練課、觀看紀錄片和訪談，也可以上網收聽各領域菁英的發言。碰到任何有趣且說服力高的語句應該馬上做筆記，然後將這些技巧應用在自己的交談中。

❸ 模稜兩可的藝術

對外交官來說，何時該交代所有事實、何時該看情形只透露部分實情，是非常重要的。誠如馬克威廉斯所言：「外交官必須精通多種說話方式，其中包括模稜兩可。」或者，像艾蜜莉狄更絲寫過的「說出真相，但委婉道來」。

❹ 抗壓性夠（就算你其實很緊張）

英國已故首相邱吉爾處理高壓危機的能力為人稱道。他的辦法是運用領袖式語言，同時以幽默來點綴外交手腕。當面臨殘暴的希特勒，邱吉爾說：「所謂姑息養奸者，就是那些不斷餵養鱷魚，希望自己是最後一個才喪命的人。」他精準的使用片語，讓文字發揮強烈效果。運用邱吉爾的技巧，你會發現其他人將更容易接納你的看法，這也是談判場合中我們最想要的結果。（請參照 Part 5 的談判技巧）

064 如何開啟一場討論

友善

- 我喜歡有來有往的討論，我們都說說自己關於「⋯」的想法吧！
  - I love a lively debate! Let's talk about [topic].

- 現在正是討論這件事最好的時候。
  - There's no time like the present to discuss this.

- 我希望你能對此發表看法。
  - I'd love to have your take on this.

- 我們就從這裡開始討論，可以嗎？
  - Let's open this up for debate, shall we?

- 我們自由發言吧！我發現這樣效果最好。
  - Please speak freely—I think we'll work better that way.

- 我希望我們能先針對某些主題集中討論。
  - I'd like for us to bat around some ideas for a moment.

- 我早就想和你討論這件事情了。
  - I've wanted to talk with you about this for a long time.

- 這件事想要成功的話，我們一定要互相討論。
  - Only by talking things through can we help one another succeed.

- 我想多聽聽、多了解你們的想法。
  - I want to listen and I want to understand.

- 我希望大家能針對「⋯」多發表些意見。
  - I'd like for us to elaborate a bit on [topic].

- 現在請大家針對這個主題自由發言吧！
  - The topic is now open for discussion.

- 我知道這會是一次很有建設性的討論。
  - I know our discussions will end up being constructive.

- 每次的討論總是可以帶來正面的結果。
  - Our discussions have always led us toward an amicable solution.

- 我不只是想討論，更希望能得出具體結果。
- I'm not looking to simply talk; I'm looking to make something happen.

- 這次討論的主題是「…」。
- The topic for this session is [topic].

- 我建議大家輪流發表對於「…」的看法。
- I suggest we all share our views about [topic].

- 我們一起深入探討「…」這個題目吧！
- Let's dive into the subject of [topic].

- 我希望今天大家能對「…」有更深入的了解。
- I'm hoping we can come to an understanding about [topic].

- 今天的討論不管得出什麼結果都是好的。
- Whatever results from our discussions will be just fine.

- 唯有徹底討論，對我們的困境才有幫助。
- Real discourse can only help us during tough times.

- 良性的對立可以幫助我們雙方成長。
- A good rivalry will only make us both stronger.

- 不管我們討論出什麼，一定都會有正面的幫助。
- Anything we discuss must lead to positive results.

- 溝通是必要的。
- Dialogue is absolutely necessary.

- 我們發表看法的時候不應該有所顧忌。
- We should all be able to express ourselves without fear.

- 我想釐清所有的疑點，並且讓反對意見消失。
- I want to eliminate all ambiguities and remove all objections.

- 我們再各自說明一次自己的論點吧！
- Let's each go over our side of the argument.

- 如果一開始就能把複雜的部分搞清楚，情況會好很多。
- We'll all feel better if we eliminate the complexities right from the start.

- 我們所有的討論都必須朝具體的方向前進。
- Any discussion we have needs to lead to something concrete.

- 首先，我們要找出目前遇到的問題。
- Let's begin this by first recognizing the problem(s) we face.

- 我希望這次的討論不會出現負面的結論。
- I hope nothing negative will come from our discussions.

- 我發現我們很難消除彼此的歧異，該怎麼解決好呢？
- I have a hard time getting past our differences—how can we fix that?

- 我很樂意進行公開討論，你呢？
- I'm okay with an open discussion—are you?

- 我們現在就講清楚吧！
- Let's hash this out right now.

- 既然我們早晚都要面對這件事，不如現在解決吧！
- Well, we have to hammer this out sooner or later, so it might as well be now.

對立
- 我們乾脆把話講清楚說明白，好嗎？
- Let's just get this out in the open, shall we?

065 如何結束一場討論

友善
- 真高興能和你討論。
- I am so glad we talked!

- 今天的對話很有收穫。
- I feel really good about today's conversation.

- 我提議大家針對剛剛的談話做個結論，然後一起去喝一杯吧！
- I vote that we wrap up the proceedings and go out for a beer.

- 我不能再聽下去，大腦收到的資訊已經超載。（玩笑口吻）
- I can't take in any more information—my brain is fried! [joking]

- 其實意見不同也沒有不好，反而更有趣。
- It's okay that we don't see eye to eye—in fact, it keeps things interesting.

- 如果再繼續聊下去，我怕我會滔滔不絕一直講下去。
- The danger of continuing our little talk is that I might never want it to end.

- 我相信所有事情都談完了，還有什麼要補充的嗎？
- I believe everything has been resolved—any final words?

- 看起來一切都挺順利的，今天就討論到這裡吧！
- Seems like we're back on track; I say we end things here.

- 或許我們可以過幾天再來討論這件事，你說呢？
- Maybe we can revisit this at a later date—what say you?

- 我很想繼續和你聊下去，可惜現在真的沒時間。
- I could listen to you for hours, but I just don't have the time.

- 和你說話真是獲益良多，真可惜我還有另一個行程。
- I learned a lot talking to you, but I have another engagement.

- 我很想繼續聊下去，但〔另一個活動〕已經遲到了。
- I'd like to continue with this, but I'm late for [task].

- 我想這次的談話沒辦法在有限的時間內結束。
- This conversation will take more time than I have right now.

- 在此宣布本次討論於〔時間〕正式結束。
- Let it read in the minutes that we closed the debate at [time].

- 謝謝，這個主題就到此告一段落。
- Thank you, but this subject is now closed.

- 已經沒有繼續討論的必要。
- There's really no reason to continue, is there?

- 這個話題一時之間很難有明確的答案。
- It's not an argument that has an easy solution.

- 眼下我們只能同意雙方暫時沒有共識。
- For now, we'll have to agree to disagree.

167

- 抱歉，對於此次討論我已經沒什麼好說的了。
- Sorry, I can no longer bring anything helpful to this debate.

- 看來這個問題比我們聰明的人也討論不出結果，就此打住吧！
- People smarter than we are don't agree on this—let's just drop it.

- 我知道繼續討論下去會有什麼結果，還是就此結束吧！
- I know the usual outcome of this kind of exchange, so I'd rather avoid it.

- 如果我們繼續討論下去，最後可能會兩敗俱傷。
- If we continue in this vein, the conversation will end badly.

- 我不認為我們應該繼續再談下去。
- My conviction that we should continue this discussion is rapidly diminishing.

- 現在結束討論對大家都好。
- To close these talks would be a blessing to everyone involved, I think.

- 再繼續講下去也沒什麼意義。
- There's really no point in continuing.

- 我們停在這裡吧！
- Let's terminate the proceedings.

- 我連一分鐘都不想再說了。
- I will not continue with this for another minute.

- 我想討論已經太過火了。
- I think we've crossed a line, here.

- 這整件事根本行不通。
- This is a dead-end subject.

- 就這樣，我要離開了。
- That's it, I'm walking out.

- 談話到此結束。
- This conversation is over.

強制

- 我要走了。
- I'm outta here!

066 如何開始一場演說

日常對話

展現外交手腕

處理衝突與憤怒

4 展現外交手腕

談笑風趣

解決問題

展現尊重與關懷

掌控勝局權的帝王課

正式

- 各位先生女士，各位嘉賓，…
- Ladies, gentlemen, and honored guests...

- 我親愛的兄弟姊妹，…
- My dear brothers and sisters...

- 我在此懷著誠摯的心…
- With solemnity in my voice...

- 在此，我備感榮幸…
- With the greatest humility...

- 感謝在場每一位嘉賓的蒞臨。
- Thank you for coming, all of you.

- 能夠有這個機會和各位說說話，我備感榮幸。
- I am filled with gratitude to be speaking with you today.

- 今天我們能聚在一起，是種難得的緣分。
- It is not by happenstance that we are all together today.

- 我謹代表〔某位嘉賓或機構〕，揭開今天的序幕。
- On behalf of [honored guest or entity], I'd like to begin by saying...

- 今天大家有幸齊聚一堂。
- This is an auspicious occasion for everyone involved.

- 首先我要說的是…
- I would like to start by saying...

- 首先我要表達對〔人名〕的尊敬與感謝。
- I will begin by offering a simple [acknowledgment/homage] to [honored person's name].

- 今天，我敞開心門站在大家面前。
- I stand here before all of you with an open mind and an honest heart.

- 人生中總是有需要謹慎發言的時候，例如現在。
- There's a time when one ought to use precisely the right words, and that time is now.

- 今天是〔人名或機構名〕的大日子。
- This is a proud day for [honored guest or entity].

- 我想先說說一個簡單的想法。
- Let me begin with a simple thought.

- 在正式開始之前，我必須先說今天真的很高興能看到各位。
- Before I begin, I would like to say that I am very happy to see all of you here today.

- 今天我有幾點想法要和各位分享，希望大家仔細聆聽。
- I've got a few points to make here today, so please bear with me.

- 我好不容易把大家都困在這裡了（玩笑口吻）。今天我打算談一談⋯
- Now that I've got you all cornered [joking], I'd like to start by saying...

- 大家好，請容我先做一段自我介紹。
- Hello, all! I will start by introducing myself.

- 我保證我的發言會簡短扼要。
- I promise this'll be short and sweet!

隨性
- 各位先生女士⋯
- Ladies and germs...

067 如何結束一場公開演說

正式
- 我今天真的很高興能夠和各位聊聊這個話題。
- I am filled with gratitude that I was able to speak with you today.

- 今天能夠在這裡演說是我的榮幸。
- It has been an honor speaking with you today.

- 謝謝大家給我這個演說的機會。
- Thank you for allowing me to speak to you today.

- 最後，我要再次感謝大家對〔公司名／專案名／目標〕的支持。
- I will end by acknowledging our mutual commitment to [company/project/cause].

・在最後，請容我再強調一次…

・In closing, allow me to reiterate one last time...

・讓我在此替各位歸納今天的內容重點。

・Allow me to close by recapping my major points.

・最後我要說的是…

・I will close by saying...

・正如羅斯福所言：「胸懷大志，腳踏實地。」

・As Theodore Roosevelt said, "Keep your eyes on the stars but keep your feet on the ground."

・我想在座各位或許有許多問題，大家現在請自由提問。

・I expect a fair number of questions, so let me turn the podium over to you.

・謝謝各位的參與，祝大家有美好的一晚／天。

・I appreciate your attention; I hope you all have a great day/night.

・各位的問題是最好的結尾，特別是那些我能夠回答的問題。（玩笑口吻）

・The best way to end is with a few questions—ones that I hope I have answers to. [joking]

・現在開放現場提問。

・And now for the Q & A.

・再美好的時光也有結束的時候。（玩笑口吻）

・All good things must come to an end. [joking]

・我想今天的演說就到此結束。

・So I guess this is the way this speech ends—not with a bang, but with a whimper. [joking]

・謝謝各位，你們是最棒的。

・Thank you, you've been great!

隨性

068 如何指控他人

委婉

- 我完全沒有冒犯你的意思，但是最近有一件事情引起我的注意。
- I truly don't mean to offend you, but something has come to my attention.

- 抱歉，但最近有些事讓我不得不找你聊聊。
- An unfortunate situation has come up that I need to talk to you about.

- 我不知道該怎麼說才好，但是…
- It's really hard for me to say this, but...

- 如果我不關心你，我就不會說出…
- If I didn't care about you, I wouldn't say this...

- 雖然我沒有資格指責任何人，但是…
- It's not my place to point fingers, but...

- 我很抱歉，但是我必須直話直說，…
- I regret that I have to be so direct, but...

- 我實在不願意指責你，但…
- It's difficult for me to accuse you of anything, but...

- 告訴你這件事情是我的職責所在。
- I would be remiss if I didn't bring this to your attention.

- 有件事我必須老實跟你說。
- There's something I need to confront you about.

- 你必須為自己的行為負責。
- You need to take responsibility for your conduct.

- 我沒有侮辱你的意思，但我知道這是你做的。
- With all due respect, I know you did this.

- 俗話說，無風不起浪。
- There's no smoke without fire.

- 現在是你面對現實的時候。
- It's time you faced the music.

・無論如何你就是不肯認錯？	・Will you never admit to your wrongs?
・你不覺得於心有愧嗎？	・Don't you feel guilty?
・這是你的錯！	・I blame you.
・承認吧！錯在你。	・You're at fault here—just admit it.
・我要揭發你。	・I'm calling you out.
・都是你的錯。	・It's all your fault.
・你從實招來吧！	・'Fess up!
・你看起來鬼鬼祟祟，一定有問題。	・If it walks like a duck..

直接

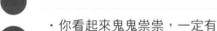

069 如何凝聚共識

溫和

・只要同心協力，問題一定能夠解決的。	・I think we can work something out if we all worked together.
・如果我們彼此都能各退一步，自然能夠互相溝通。	・Compromise can form a bridge between our differences.
・只要大家團結，沒有解決不了的問題。	・With solidarity, we can conquer any problem.
・眼前的問題如此棘手，找出共識是當務之急。	・The question is a tough one; all the better to form a consensus.
・只要我們目標一致，就能夠成功。	・We'll succeed if we find common ground.
・相信我們可以達成共識的。	・I am positive we can come to an accord.

- 我知道我們能夠找出一個折衷的方案。
- I know that we can come to a compromise.

- 現在正是我們團結起來發揮力量的時刻。
- It's time that we all came together and made this happen.

- 我們現在最需要的就是凝聚共識。
- It's vital that we forge a consensus.

- 我想大家都同意一個共同的對策是必要的。
- I think we'd all agree that we need a unified solution.

- 問題雖然棘手，但只要大家合作就能順利解決。
- This issue is complex, but good things will happen if we come to a consensus.

- 合作是現在唯一的辦法。
- Collaboration is the only way forward at this time.

- 我們大家最終的目的都是一樣的。
- At the end of the day our needs are all the same.

- 我願意討論各種可能性，你呢？
- My options are completely open—are yours?

- 我想大家應該都各退一步。
- It's time to make some concessions.

- 我們現在一起來找出共同的立場吧！
- Let's seek an accord right now.

- 沒有必要完全順著某一方的意見。
- There's no need for complete capitulation on either side.

- 在達成協議之前，我們必須先對事情有相同的了解。
- We need to get on the same page before coming to an agreement.

- 或許大家換個角度想，就有可能達成共識。
- Try to see it another way so that we can come to an agreement.

- 要解決這個問題，大家都必須稍微妥協才行。
- To solve this problem, everyone involved needs to be willing to give in a little.

· 為什麼我們不能站在同一個陣線上呢？	· Why can't we all get on the same page?
· 如果最後達不成共識，對誰都沒有好處。	· If no agreement comes out of this, nobody benefits.
· 再不能達成協議，我們的麻煩就大了。	· Either we come to an accord, or we have a serious problem.
強勢 · 爭到最後只會兩敗俱傷。	· Nobody wins if everyone loses.

070 如何處理意見不同的場面

接受 · 如果大家想的都一樣，那也沒有溝通的必要了，對吧？（玩笑口吻）	· If we all agreed on everything, what would be the point of talking? [joking]
· 我們顯然意見不同，不過這樣也無妨。	· We obviously have a difference of opinion, and that's okay.
· 我完全了解你為什麼會這樣想，請再多說些你的想法吧！	· I certainly understand why you would think that—tell me more.
· 你似乎對整件事非常了解，但是你怎麼能這麼肯定呢？	· You seem to know a lot about this—but how can you be so sure?
· 這件事或許還有討論商量的空間。	· There's probably room for some interpretation/nuance here.
· 我相信大家過些時候就能達成協議了。	· I'm sure we will come to an agreement in time.
· 就算我們對很多事的看法不同，但不代表我們不能同意某些事情。	· Just because we don't agree on everything doesn't mean that we can't agree on something.

- 我們雙方彼此都有些顧慮，不如試著往前邁進吧！
- Both of us clearly have misgivings, but let's try to move forward.

- 我了解你的意思，但我的看法和你不一樣。
- I understand how you see it; I just see it differently.

- 我們有點雞同鴨講，大家試著多用點心溝通吧！
- We're not speaking the same language yet—let's try harder.

- 你覺得我們可以找出折衷的辦法嗎？
- Do you think we could meet somewhere in the middle on this?

- 照你看來，我們該如何達成共識？
- In your opinion, how can we best reach common ground?

- 我們不要因為小小的意見不同就整個停擺。
- Let's not get bogged down by a small difference of opinion.

- 該怎麼做大家才能夠達成協議呢？
- What do we need to do to come to an agreement?

- 請恕我直言。
- If I may react/respond to that.

- 我想要說的是…
- I would like to argue that...

- 你是不是至少同意…？
- Do you at least agree that...?

- 我知道很多人對這件事的想法都和你不同。
- I know many people who would think differently.

- 這個爭議已經延續好幾百年了。
- This subject has been debated for hundreds of years.

- 不是每一個意見都有一樣的價值。
- Not all opinions are equal.

- 很明顯大家意見分歧。
- There's clearly a difference of opinion, here.

- 每個人的意見都一樣重要。
- No-one's opinion is priviledged.

- 我對這件事的解讀不同。
- That's not my understanding of the issue.

- 你或許應該重新思考一下你的意見。
- You might want to reconsider your opinion on this.

- 相信你也看得出來，我很難接受你的看法。
- As you know, I have a hard time subscribing to that point of view.

- 所謂的事實都是相對，而非絕對。
- The truth of any situation is relative.

- 先證明你所說的吧！
- Try to realize what you're saying here.

- 你不覺得把事情講得太誇張了嗎？
- Don't you think you're overstating the issue a bit?

- 我不懂你該怎麼自圓其說。
- I don't see how you can rationalize/justify your comments/beliefs.

- 很明顯我們的看法不同。
- It's obvious we don't see eye to eye.

- 既然我們意見不合，也無需再多言。
- There's no point in discussing this any further since we don't agree.

- 既然談了半天還是在原地打轉，不如就此結束。
- We're nowhere even close to an agreement, so we should just call it a night.

- 如果我們無法達成協議，那就沒有談下去的必要。
- If we can't come to an agreement, there's no point in continuing our dialog.

- 別說些自己都不了解的話。
- Don't speak about what you don't know.

- 我沒心情聽你胡說八道。
- I'm not in the mood to tolerate such nonsense.

- 你的說法根本就是旁門左道。
- Your convictions are heretical.

- 拜託你別再癡人說夢了。
- Come back down to planet Earth!

拒絕

177

071 如何避免敏感話題

禮貌

- 或許我們應該先把這件事放在一旁。
- Maybe it's best to leave that alone for now.

- 我們換個話題，維持歡樂氣氛吧！（玩笑口吻）
- Let's change subject and remain blissful. [joking]

- 相信我，討論這件事只是自找麻煩而已。（玩笑口吻）
- Believe me, this topic is nothing but trouble. [joking]

- 你知道什麼是「潘朵拉的盒子」吧？（玩笑口吻）
- You know what they say about Pandora's Box. [joking]

- 我寧可跳進油鍋也不想談這件事。（玩笑口吻）
- I'd rather be boiled alive than talk about that. [joking]

- 這已經是個敏感的話題了，你知道吧？
- This has become kind of a sensitive subject, you know?

- 我們還是不要談這件尷尬的事吧！
- Let's not open a can of worms.

- 我們先把正事談完吧！
- Let's stick to the matter at hand, shall we?

- 我們應該保持現在正面的氣氛，別去談那件事。
- Let's keep this positive vibe going and not talk about that.

- 我們可以講點別的事嗎？
- Can we consider discussing something else?

- 或許我們現在暫時不要去碰這個話題吧！
- Maybe we shouldn't touch on that topic just now.

- 我們下次再來談這件事吧！
- Let's table this discussion for another time.

- 如果你不介意，我現在不想談這件事。
- I'd rather not talk about that, if that's okay with you.

- 一直談這件事或許有損我們的專業。
- It probably would be unprofessional to proceed along those lines.

- 如果再不換個話題，我們可能會有麻煩。
- If we don't change the subject, I can see trouble brewing.

- 這件事已經談過了，我不想再講一次。
- We've talked this through already so let's not go there again.

- 我不想引起衝突，所以請略過那件事不談，好嗎？
- I hate confrontation, so let's just avoid that whole topic, okay?

- 這個話題讓我不太自在，可以先擱在一邊嗎？
- This is getting a little uncomfortable. Can we let it slide for a little while?

- 無論對誰，這都不是個很好的話題。
- This is never a good topic of conversation—for anyone.

- 我們可以先不談那件事嗎？至少等我離開再談。
- Can we move away from this topic—at least until I leave?

- 每次談到這件事就火藥味十足，而且一次比一次嚴重。
- I've heard this tune before, and it gets more discordant every time.

- 上次談論這件事時整整花了一個禮拜／月才平息所有的糾紛。
- The last time we discussed that, it took a [week/fortnight/month of Sundays] to calm everyone down.

- 這是個禁忌的話題。
- That topic is taboo around here.

- 我們時間有限，先略過這個話題吧！
- We don't have much time, so let's just keep it light.

- 你覺得談這個真的對大家有幫助嗎？
- Do you think this discussion will bring us any closer?

- 繼續談下去也是徒勞無功。
- Pursuing this dialog any further is pointless.

- 你知不知道這個話題對我來說是多大的麻煩？
- Don't you realize how much of a problem this topic is for me?

- 這個話題就像是無間地獄，我們換個正面一點的吧！
- That topic is five miles of bad road. Let's pick something more productive.

- 這樣講下去，場面只會越來越僵。
- Things only get more tense with that kind of talk.

- 一直停留在這個話題，只會讓人氣到吐血。
- Ruminating on this will only lead to heartache, or worse.

- 談這樣的話題根本就是死路一條。
- It's conversational suicide to even consider a topic like that.

- 老天，又來了。
- Oh boy, here we go again.

- 我不想談這件事，到此為止。
- I don't want to talk about it, end of story.

強硬

- 我不會和你談這件事的。
- I will not engage with you about this.

072 如何做建議

溫和

- 我有些想法，你們不妨聽聽。
- I have some thoughts but I want to tread lightly.

- 集思廣益總是好的，要不要聽聽我的看法。
- There is wisdom in the counsel of many—would you like to hear my thoughts?

- 我們應該是好朋友吧？你覺得〔建議內容〕怎麼樣？
- We're pretty good friends, right? What would you think of [suggestion]?

- 我有些主意，有興趣聽嗎？
- I have some ideas—would you like to hear them?

- 我們都這麼熟了，我想我可以告訴你我的想法。
- We've gotten to know each other so well, I think we can get personal.

- 我只是想把想法說出來，如果…，你覺得如何？
- I'm just thinking out loud, but what if you...?

- 雖然這只是我個人的想法啦！但是…
- This is only my opinion, but...

- 我可以實話實說嗎？
- May I be honest with you?

- 我可以老實說嗎？
- May I speak frankly?

- 你介不介意我提出一個建議？
- Do you mind if I offer you a suggestion?

- 希望你別誤會，但是…
- Don't take this the wrong way, but...

- 或許你可以…
- It might be a good idea if you...

- 我不想讓你感覺不舒服，但是…
- I'm not trying to make you feel badly, but...

- 我沒有冒犯你的意思，但是…
- I don't mean any disrespect, but...

・我希望你別誤解了我的建議。

・I hope my suggestion doesn't come across the wrong way.

・希望你不會覺得我太無禮，但或許你可以考慮…

・I hope you don't find this offensive, but perhaps you should consider [suggestion].

・你不見得要採納，但我有些建議可以參考一下。

・You don't have to take it, but here's some good advice.

・我會建議你…

・I would recommend that you...

・這件事情上，我想給你一些有用的建議。

・Let me give you some valuable advice about this.

・你現在最好的做法就是…

・The best thing for you to do at this point would be to...

・這可能對你有幫助。

・This might help you out.

・我想告訴你…

・Let me just say this...

・一個小建議，…

・A word of advice...

・看來你需要一些客觀的意見。

・Seems like you could use an objective opinion.

・你需要清醒清醒，不要一錯再錯。

・You could not be more wrong—you need to listen to reason.

直接 ・你完蛋了。

・It's your funeral.

如何回覆不想採納的建議

- 謝謝你對我的生活如此關心。
- Thank you for caring enough to speak into my life.

- 謝謝你的好意，我很感激你這麼誠實。
- It's very kind of you and I really appreciate your honesty.

- 喔，這個看法很特別，謝謝。
- Wow, I never saw it that way. Thanks!

- 我一定會把你的建議納入考慮的。
- I will certainly take that into consideration.

- 你的話非常發人省思，謝謝。
- You gave me a lot to think about, thank you.

- 我很高興你特地花時間和我說這些。
- I'm really pleased you took the time to tell me these things.

- 謝謝你告訴我。
- Thank you so much for letting me know about this.

- 如果你還有其他建議，請務必告訴我。
- If you have any other suggestions, please let me know.

- 謝謝你指出這些部分。
- Thanks for pointing that out.

- 你的建議很好。
- That's good to know.

- 我一向喜歡聽取他人的意見。
- I always like to hear others' opinions.

- 你想提出任何建議都行。
- I'm open to whatever you might suggest.

- 謝謝，我很感謝你的關心。
- Thanks, I appreciate your concern.

- 謝謝你的幫忙。
- Thanks for wanting to help.

- 我知道你是為我好，但真的不用這麼擔心。
- It's very nice for you to worry about me, but you really shouldn't.

‧ 我知道你是好意幫我，但我對現狀很滿意。
‧ You're very kind to try to help me, but I'm happy like this.

‧ 我了解你的意思，不過我還是要照原訂的計畫進行。
‧ I get what you're saying, but I think I'm okay with my plan.

‧ 我知道你是一番好意，但我想自己處理這件事。
‧ I know you mean well, but I prefer to keep my own counsel on the matter.

‧ 你提出這些建議／意見的根據是什麼？
‧ What experience/knowledge do you base your advice on?

‧ 謝謝你的關心，但我很好。
‧ Thanks for your concern, but I'm doing just fine.

‧ 這件事情我會自己決定。
‧ I will decide to take my own action on the matter.

‧ 我完全知道自己在做什麼，謝謝。（諷刺語氣）
‧ I'm fully aware of what I'm doing, thanks. [sarcasm]

‧ 我的事會自己處理，不需要別人插手。
‧ I don't need anyone to look out for my interests but myself.

‧ 你的意見根本一文不值。
‧ Your advice is worth about what I paid for it.

強硬
‧ 如果我需要你的意見，我自然會問。
‧ If I wanted your opinion/advice, I would have asked for it.

當有人為了某件事指責你

禮貌

・啊！好像在開庭審理。（玩笑口吻）

・Ah, court is in session! [joking]

・我不知道你怎麼會得出這個說法，請你詳細解釋。

・I'm not sure how you arrived at that conclusion, but I want to hear you out.

・請讓我和你一起查清楚真相到底是什麼。

・Please let me help you find out what really happened.

・我很難過原來你對我的看法是這樣，要怎麼做才能改變你的想法呢？

・I'm hurt that you would even think that—what can I do to convince you otherwise?

・就算我想承認也沒有辦法，因為根本就不是我。

・I wish I could say I did it, but that would be a lie.

・我了解為什麼你會這樣想，但事情根本不是那樣的。

・I understand why you would think that, but it's simply not the case.

・抱歉，我不是很了解你的意思。

・I'm afraid I'm not following you at all.

・你說的根本與事實完全不符。

・You can't be any further from the truth.

・你要為你的指控提出證據。

・The burden of proof is on the accuser.

・我不接受你的指控。

・I'm going to have to defend myself, here

・對事情不夠了解的人很容易自己妄下結論。

・It's very easy to assume when we don't know.

・我完全不知道你在說什麼。

・I have no idea what you're talking about.

・別怪我，這件事與我無關。

・Don't blame me—it wasn't my responsibility.

- 你根本是胡說八道。
- You're speaking completely out of context.

- 我無法接受你說這些顛倒是非的話。
- I will not allow you to say things that are inaccurate.

- 如果你知道事情真相，就應該明白根本不是我的錯。
- If you knew the facts, you'd see that it's not my fault.

- 在指責別人之前，你應該先把事情弄清楚。
- You should be sure before you start accusing others.

- 我覺得你根本沒說到重點。
- I don't think we're getting to the heart of the issue.

- 我沒時間回答你這些指控。
- I have other things to do than answering to polemics.

- 在指控別人之前，應該先給對方說話的機會。
- Before you point a finger at someone, you should give him/her a chance to speak.

- 問題遠比你說的複雜許多。
- The problem is much more complex than that.

- 指責別人並不能解決問題。
- Pointing fingers doesn't help anything.

- 我不承認也不否認。
- I will neither confirm nor deny that.

- 如果你自己真的那麼完美，再來指責別人吧！
- Let he who is without sin cast the first stone.

- 別五十步笑百步了。
- People who live in glass houses shouldn't throw stones.

- 很明顯，你的理解有問題，那根本不是我的錯。
- Clearly something went terribly wrong, but it wasn't my fault.

- 請你明白這完全不是我的本意。
- Please understand this was never my intention.

・這不是任何人的錯，到處指控別人一點意義也沒有。 ・ It's nobody's fault, so it's useless to assign blame.

・看來我只是個代罪羔羊。 ・ Seems that I'm a victim of circumstance.

・你怎麼會有這麼離譜的推論呢？ ・ How did you come to this erroneous conclusion?

・我想知道你從哪裡得到這些想法的？ ・ I'd like to know who or what is your source?

・不是說「在證明有罪之前，人人都是清白的」嗎？ ・ Whatever happened to "innocent until proven guilty"?

・這件事的發生和我完全沒有關係。 ・ This didn't happen on my watch!

・就算我的錯吧！（諷刺語氣） ・ Mea culpa. [sarcasm]

強硬 ・我要告你毀謗。 ・ I will sue you for slander/libel.

展現外交手腕 4

075 當你是謠言主角

禮貌

- 我很抱歉，但是我不回應這些不實的流言蜚語。
- I'm sorry, but I never comment on half-truths and innuendoes.

- 哈，我還真希望我的生活真的有這麼精彩。（玩笑口吻）
- I only wish my life were that interesting! [joking]

- 這件事目前還在調查中，恕我無法回應。
- This is an ongoing investigation/problem, so unfortunately I can't comment on it.

- 目前沒有任何進一步消息，我也不做任何聲明。
- I don't have any further information and will not release any statements at this time.

- 世界上還有很多比這個更重要的事。
- There are more important things going on in the world than this.

- 這完全誤解了我說的話／做的事。
- That is a misinterpretation of what I actually said/did.

- 我奉勸大家對謠言別太當真。
- My advice on these rumors is to let them go.

- 謠言和影射就只是謠言和影射罷了！
- Rumor and innuendo, that's all it is

- 謠言並不是事實。
- Rumors are not the best sources of facts.

- 這些都是道聽途說的謠言，我不回應。
- This is just rampant insinuation and I have nothing to say.

- 我不會談這件事。
- I'm not going to discuss that.

- 回應這些不實謠言對我沒有任何幫助。
- It would be unwise for me to respond to such nonsense.

・我們不要繼續以訛傳訛，鼓勵這種幼稚的造謠行為。
・Let's not perpetuate such infantile behavior by giving it lip service.

・我不想對這個八卦再做回應，但我要說…
・I didn't want to give this gossip any attention, but I must say something...

・這種八卦很快就會消失。
・Such gossip will not be tolerated much longer.

・這些謠言根本不值得做任何回應。
・These are half-baked lies that don't dignify a response.

・我沒必要而且也不會回答你。
・I don't report to you and will not respond.

・什麼時候開始，我要向你報告了？
・Since when am I accountable to you?

強硬

・那又如何？
・So what?

Part

談判時刻

一般人聽到「談判」兩個字，都會不由自主得肅然起敬。掌握住幾個技巧，就會發現談判並不如字面上那麼困難，甚至讓其他人完全無法忽視你。

溝通哪些事

- 如何詢問問題
- 想拖延回答時間時
- 如何強調重點
- 想說服對方時
- 如何接受提議
- 打算拒絕對方提議時
- 如何向他人推銷

- 如何討價還價
- 當陷入僵局時
- 如何要求對方妥協

> 談判是要得到對手可以提供的最好條件
> ——馬文蓋依（美國音樂家）

　　知道如何談判是一個好領導者最重要的技巧之一。懂得談判的人通常可以獲得高薪，事實上談判並不如字面上那麼困難。不只是公司的業務還有採購，任何需要與人接觸的職位都需要和人談判。新婚夫妻需要協調彼此的生活空間；老師在學校上課也需要用到談判；銀行家想要拿到任何生意非得談判；甚至連音樂家也必須要為了自己的演出費用而談判。如果你不知道或是害怕談判，就容易有被忽視的危機。下面幾個要點能夠改進你的談判技巧。（請注意，有些部分和上一章「展現外交手腕」有所重疊）

❶ 獲得他人的信任

　　談判技巧高的領導者通常知道如何可以很快獲得各方的信任。獲得信任的方法之一是適度提供自己的談判目的資訊，但也不能太多。當對方知道你想要什麼，自然會更信任你。

❷ 別解釋太多

　　這和前面的提供資訊剛好相反；對手越了解你的目標，自然就越能避免同意甚至拒絕你。有時候保持莫測高深的形象反而對你有利。

談判過程中，會讓自己居於劣勢的資訊要盡量藏好。同時，透漏自己意圖的時候用字力求簡潔，說完必要的內容就停止，適時保持靜默。替自己解釋太多只會顯得你居於劣勢或者決心不夠，甚至看起來充滿戒備。

③ 態度從容自在

不懂得談判的人總是讓對手看出自己的不足，這絕對要避免。不管心裡有多激動，外表都應該保持鎮定，喜怒不形於色。先從生活中的小談判開始練習，遇到大場面自然能夠進步。

④ 隨時準備離場

通常，最不在乎談判結果的人，或者看起來最不在乎談判結果的人，說話最有分量。既然你掌握了大局，早就規劃好談判的劇本，包括你該離場的時候。

⑤ 沉默是金

你不需要從頭說到尾，有時候談判中短暫的沉默甚至暫停，反而能發揮更大的效果。沉默通常也會讓你的談判對手感到不自在，可能因此亂了陣腳，做出更大的讓步。

076 如何詢問問題

正式

- 請問你願意幫忙回答一個問題嗎？
- Would you be so kind as to answer this one question?

- 如果你不介意，我有個問題想請教。
- I do have one question, if you don't mind.

- 如果可以，我想提出一個重要的問題。
- I'd like to raise an important point, if I may.

- 不好意思，或許剛剛你已經提過，但是…
- I apologise if we've covered this before, but....

- 我希望能對這件事有更深入的了解。
- I am seeking to understand this matter more fully.

- 我的問題或許有點繁複，但是想請教…
- While it may sound rhetorical, my question is...

- 另一個有待釐清的問題是…
- Another question remains to be clarified...

- 我希望能聽聽你對這個問題的回答。
- I'd like to hear your solution to this.

- 請教你一個簡單的問題。
- Here's a simple question.

- 我就不再拐彎抹角，直接問你了。
- I'll go the direct route and simply ask this question.

- 我有事要問你。
- I have something to ask.

- 也許你會知道這個問題的答案。
- You probably know the answer to this.

- 我一直在思考一個問題。
- Here's something I've been wondering about.

- 我在想，如果…
- I was wondering if...

・有人知道答案嗎？	・Anyone know the answer to this?
・我已經黔驢技窮了，有人能夠幫幫我嗎？	・I'm stumped—can someone help me out here?
・有好奇心的人都應該會想知道吧！	・Inquiring minds want to know.

隨性

077 想拖延回答時間時

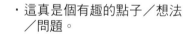

專業 ・這是個好問題，我必須好好思考過才能回答。	・That's a very good question—allow me to think before I respond.
・這真是個有趣的點子／想法／問題。	・What an interesting idea/thought/question.
・請讓我在回答你之前花幾分鐘整理一下事情的始末。	・Please allow me a moment to ponder the ramifications before I speak.
・我需要一些時間來整理一下思緒。	・I will need to gather my thoughts here for a moment.
・這個問題需要更深入的研究。	・That's going to require more research.
・回答問題之前我需要收集更多資訊。	・I need to gather all the facts before I weigh in.
・這的確是個大家都想知道的問題。	・That's on a lot of peoples' minds these days.
・我很高興你提到這件事。	・I am really glad you brought that up.
・你總是有很精彩的論點。	・You always make such excellent points.
・你的問題非常合理，的確需要一個好的解釋。	・Your inquiry is a legitimate one and calls for a qualified response.

- 你提出的論點很有趣，我需要一些時間想想。
- I think you are raising a very interesting point.

- 這是個好問題，值得好好思索一下。
- That's a good question and it deserves a clear answer.

- 我先簡短地回答你，以後我們再深入討論。
- I'll give you the short answer; we can talk more indepth later on.

- 由於時間／背景資料有些變動，讓我想一想。
- Let me think about it, as times/the facts have changed.

- 這需要大量的專業知識判斷。
- That takes a good deal of specialized knowledge.

- 讓我稍微想一下。
- Let me think on it for a second.

- 我不記得有聽過這件事。
- I don't recall hearing that before.

- 你可以解釋或是重講一遍你的問題／觀點嗎？
- Can you clarify or restate the question/your point?

- 我不確定你問這個問題的目的是？
- I'm not sure I understand where you're going with that.

- 你為何這麼問？
- Why do you ask?

- 天曉得？
- Who knows?

隨性
- 我真的是無話可說了。
- I honestly don't even know what to say.

客氣

・請容我強調這件事的重要性。	・Allow me to stress just how important this is.
・如果可以的話，我想特別強調這件事。	・I would like to emphasize this point, if I may.
・我實在無法形容這件事有多麼重要！	・I cannot stress enough just how critical this is.
・精準的言語是最重要的。	・Precision of language is of utmost importance, here.
・請特別記下這一點。	・Please take especial note of this point.
・我真的得要再強調一件事，那就是…	・I really want to emphasize the fact that...
・你可以轉述我對…的發言。	・And you can quote me on this...
・接下來我所說的事非常重要。	・What I have to say next is very important.
・雖然這件事情各家的說法不一，我的看法是…	・While there are many opinions on this, here is my take.
・我的底線是…	・Here's the bottom line.
・讓我把話講清楚。	・Let me be clear about this.
・我試著用正確的字句再說一次。	・Let me pick my words very carefully, here.
・我要再解釋一次這個重點。	・Let me make this point even more clear.
・事情的重點來了。	・Here is where the rubber meets the road.

- 這件事情並沒有所謂的「如果」、「而且」，或「可是」。
- There are no ifs, ands, or buts about this.

- 真正的重點其實只是…
- It really just comes down to this...

- 最重要／最有關聯的事實是…
- The fact that is most relevant/important is....

- 有件重要的事情是…
- An important thing to realize is...

- 我可以斬釘截鐵的說…
- I can say unequivocally that...

- 讓我換個方法再說一次。
- Let me put it another way.

- 我想我們都應該要了解…
- I think that it is important to understand that...

- 以上帝為證，…
- As God is my witness....

- 讓我來進一步解釋。
- I'll take it one step further.

- 我就不拐彎抹角，直接說了。
- I'm not going to beat around the bush, here.

- 之前就說過了，我現在再說一次。
- I've said it before and I'll say it again...

- 你應該要坐下來抄筆記的。
- You might want to sit up and take note of this.

莽撞
- 請注意！
- Now pay attention!

想說服對方時

 禮貌

- 我要怎麼做或怎麼說才能說服你呢？
- What can I do or say to persuade you?

- 請容我解釋這件事情。
- Please allow me to plead my case.

- 有很多的研究都支持這個看法。舉例來說，…
- There's a lot of research that supports this; for example...

- 任何知道這件事情的人都會覺得…
- Anyone who knows about this will say that...

- 我的資歷和經驗告訴我，…
- Time and experience have shown me that...

- 專家們同意…
- Experts would agree that...

- 請給我幾分鐘，我保證你會改變你的想法。
- Give me a few minutes and I promise you, you'll change your mind.

- 我希望你能夠試著從另一個角度來看這件事情。
- I'd like to encourage you to look at the other side for a moment.

- 只要有充足的時間，我相信能夠說服你的。
- Given enough time, I know that I can win you over.

- 就算是這個領域最頂尖的人物也會同意我說…
- The best in the field will agree with me that...

- 只要你願意聽我說所有的事實證據，相信你會改變心意。
- I believe I can sway you if you'll just listen to the facts.

- 以你的智慧一定能了解我所說的都是真的。
- Your logic must be able to discern the veracity of what I'm saying.

- 我只希望你能再重新考慮這件事。
- All I ask is that you revisit the issue.

- 事實真相會讓你清醒。
- The truth will set you free.

5 談判時刻

199

- 我不能控制你的想法，但是… · I have no right to hijack your conscience, but...

- 你絕對有足夠的智慧看出真相。 · You're smart enough to know the truth when you see it.

- 我還沒有打動你嗎？ · Am I not swaying your opinion?

- 如果你睜開眼看清楚事實，相信你會改變主意的。 · If you opened your eyes to the facts, I think you would understand.

- 這件事就像太陽從東邊出來一樣簡單正確。 · It's as plain as the nose on your face.

- 為什麼你不能理解一下我的看法呢？ · How can you not see my point?

- 任何有常識的人都會了解這件事。 · Anyone with an ounce of common sense would know that.

粗魯
- 別這麼頑固，至少先聽我講完嘛！ · Don't be so obdurate and just listen for a moment!

080 如何接受提議

正式
- 我由衷地接受這項提議。 · I am gratified to say that I accept without reservations.

- 接受這項提議，我感到既榮幸又惶恐。 · I accept, with honor and humility.

- 我無法形容這對我來說是多大的榮幸。 · I can't tell you what an honor this is for me.

- 這對我們來說都是值得驕傲的一刻。 · This is a proud moment for both of us.

- 我真的很感謝你邀請我參與。
- I'm touched to know that you made me a part of this.

- 我願意加入。
- I'm on board with that.

- 我收到了你的提議，也願意接受這個提議。
- I hear you and agree.

- 我完全同意你的提議。
- I'm with you all the way.

- 這也是我唯一想做的。
- There's nothing I would rather do.

- 我們一起來實現它吧！
- Let's make it happen.

- 我們一言為定。
- Let's shake hands on it.

- 我們開始行動吧！
- Let's pull the trigger on this.

- 記得把我寫在同意名單上。
- Sign me up on the dotted line.

- 當然囉！
- That's a no-brainer.

- 我一萬個願意。
- My answer is a big yes.

- 算我一份。
- Count me in.

隨性
- 沒問題。
- Fine by me!

081 打算拒絕對方提議時

客氣

- 我很尊敬你，但這次的提議必須拒絕。
- I must respectfully decline at this time.

- 我很抱歉這次必須拒絕你的提案。
- I'm sorry to have to say no to the generous proposition you've made.

- 我很抱歉目前無法接受你的提議。
- I apologize that I am unable to accept your offer right now.

- 很遺憾，這次我無法接受你的提議。
- Unfortunately I cannot accept your excellent proposal at this time.

- 我很感謝你的建議，但還是得採取另一個方式。
- I appreciate your suggestion, but we must find another way.

- 我必須忍痛拒絕你。
- I must decline with humility.

- 我已經接受了另一個類似的提議，所以無法接受你的。
- I have already accepted a similar proposal, so I must say no.

- 雖然我很想答應，但我想我的上司／另一半／父母是不會同意的。
- While I would love to say yes, I don't believe I could get it past my [boss/spouse/parents].

- 我不認為現在一起合作是個好主意。
- I don't think it would be wise for us to go forward at this time.

- 如果換到更好的環境，這件事有可能成功。
- This could have worked out under better circumstances.

- 抱歉，我沒有決定權。
- The decision is out of my control, sorry.

- 我不認為你的提案適合目前的情況。
- I don't think your plan is the right one for this situation.

- 抱歉，這件事情我沒辦法和你配合。
- I can't work with you on that, sorry.

・這次不行，但還是很感謝你。	・Not at this time; thanks, though.
・我們現在真的經不起任何風險／支出／障礙。	・We just can't afford the risk/expenditure/complication right now.
・你的精神令人佩服，但我沒辦法同意這件事。	・Your tenaciousness is impressive, but I just can't move on this.
・你的提案風險很大，我沒辦法接受。	・Your solution is risky; I just can't run with it.
・你的建議反而會帶來更多的問題。	・Your idea raises more questions than it answers.
・這件事行不通。	・It's a no-go.
・這件事不可能。	・It's simply out of the question.
・我不認為可行。	・I don't think it's going to work.
・你別胡說八道了。	・I'm going to have to put the kibosh on this.
・除非你付錢給我。（玩笑口吻）	・Only if you pay me for it! [joking]
・不行。	・No can do.
・我不接受。	・That's not gonna work for me.
・如果太陽打從西邊出來，我就同意。（玩笑口吻）	・Maybe when pigs fly! [joking]

直接

082 如何向他人推銷

強勢

- 如果你再不行動，保證會後悔的！
  - If you don't move on this now you'll regret it.

- 再不買你就要錯失良機了。
  - You'd better buy now before you miss out.

- 千萬別錯過這麼好的機會。
  - Don't let this be the one that got away.

- 最好趁還有貨的時候先買。
  - Get it now before it's gone.

- 今天就帶回家吧！保證不會後悔的。
  - Take it home today—you won't regret it.

- 促銷價格只有今天，明天就沒了。
  - This is a one day sale; by tomorrow it will be gone.

- 明天再來，你就晚了一步，價格更高。
  - Tomorrow you'll be a day late and a dollar short.

- 這絕對是有史以來最划算的價格。
  - This is the deal of the century.

- 這麼優惠的價格只到〔某日期〕。
  - This special price is only good until [date].

- 下次再有這麼漂亮的優惠價就要等到〔某日期〕。
  - There won't be another sale like this until [date].

- 很多人都有興趣，所以你最好動作快一點。
  - There's a lot of interest in this, so you'd better move on it.

- 今天你買到賺到。
  - This is your lucky day!

- 這種價格不買真的太說不過去了。
  - The deal is almost irresistible.

- 你買到的是一份品質。
  - You're buying quality.

・這個價格很快就要調整了。	・ It won't stay at that price for very long.
・我自己也有一個，完全改變了我的生活。	・ I have one and it changed my life.
・這是你應得的。	・ You deserve it
・你值得享受好東西。	・ You're worth it.
・你的同學／同事／朋友都會羨慕你。	・ You'll be the envy of your peers/colleagues/friends.
・下這個決定，你穩賺不賠。	・ What would it take for you to decide today?
・這是極品中的極品啊！	・ It's the crème de la crème.
・這個價格絕對划算。	・ This is a great value/bargain.
・高品質、低價位，不會有更好的組合了。	・ Quality and price—you can't ask for more.
・請務必考慮，你會發現這是你做過最好的決定。	・ Please consider it—you'll be glad you did.
・我想你會喜歡的。	・ I think you'll be happy with it.
・這對你絕對只有好處。	・ What have you got to lose?
・其實，錢再賺就有了。	・ It's only money, after all.
・買了對你也不會有任何壞處啊！	・ What's the worst that could happen?
・你隨時可以退貨。	・ You can always return it.
・不嘗試你永遠不會知道。	・ Nothing ventured, nothing gained.
溫和 ・最後的決定權在你。	・ Only you can make that decision.

083 如何討價還價

正面		
· 我相信我們可以雙贏。	· I'm sure we can get to a win-win, here.
· 我知道我們可以達成共識。	· I know we can work something out.
· 這聽起來好多了，我們再多討論討論吧！	· That sounds better—let's talk some more.
· 就這樣敲定吧！	· Let's make a deal.
· 不如我們各退一步，事情就成了。	· Let's each go halfway and we'll be done.
· 我前幾天才看到一個更高／更低的價錢。	· I saw a lower/higher price on that just the other day.
· 唯一的辦法是，你跟我各退一步。	· Meet me in the middle; it's the only answer.
· 我們都希望能成交，那就拿出成交的決心吧！	· We both want this deal to work out, so let's make it happen.
· 你不能再把價錢提高一些嗎？我都已經做出讓步了。	· Can't you budge a bit more on the price? After all, I've made concessions, too.
· 我們都希望這筆買賣能夠談成，是吧？	· We both want a successful outcome here, right?
· 買賣成不成，其實對我沒什麼影響。	· I have no problem walking away from this.
· 時間有限啊！	· The clock is ticking.
· 你要拿出更多誠意才行。	· You'll have to do better than that.
· 再幫我一點吧！	· Help me out, here.
· 我希望你能拿出更好的條件。	· I expected a better offer from you.

日常用語

辦公室情境

理論實務想

談判文字訓

5 談判時刻

解決問題

表現層面時期

不接受問題的態度

· 這完全是在商言商。	· It's not personal, it's business.
· 我們都是在討生活。	· We're both trying to make a living, here.
· 你的競爭對手看到現在這種情況，一定暗自竊喜吧！	· Your competitor is going to be happy about this.
· 你這麼堅持，真的對你會有好處嗎？	· Do you think your rigidity is serving you?
· 如果談不攏，我只好放棄了。	· I'll have to walk if we can't come to terms.
· 我一點利潤也沒有了。	· I'm not made of money.
· 哇，你的條件太侮辱人了吧！	· Wow, that's insulting.
反面 · （沉默）	· [silence]

084 當陷入僵局時

正面 · 我知道你和我一樣想解決這件事，所以你打算怎麼做呢？	· I know you want this as much as I do—what can you do for me?
· 我們差一步就能達成共識了。	· I feel that we are one step away from shaking hands.
· 我可以在這些部分妥協，所以你能夠讓步多少？	· Here are my concessions—what are you willing to bend on?
· 我已經讓步，現在換你了。	· I've given a bit, now it's your turn.
· 我們該怎麼突破目前的僵局往前邁進呢？	· How will we proceed with this two-sided quandary?

- 我們似乎陷入了僵局，你有什麼想法嗎？
- It seems we're at a standoff—do you have any ideas?

- 我們現在陷入僵局，該怎麼辦呢？
- We're in a deadlock. What should we do?

- 我不知道要怎樣脫離目前僵持不下的局面。
- I don't know how we're going to get out of this.

- 我們至少可以同意某件事情嗎？
- Can't we just agree on something?

- 現在我們雙方都無計可施，你有什麼主意？
- Both sides are in a stalemate—what do you suggest?

- 你有任何解決辦法嗎？
- Do you see a way out of this?

- 很遺憾，我們走進了一條死胡同，兩邊都動彈不得。
- Unfortunately, we've reached a dead-end and nobody is budging.

- 我不想繼續這樣你來我往，我相信你也是。
- I can't stand this constant back and forth, and I know you can't, either.

- 我希望我們能夠解開目前的僵局。
- I wish we could get out of this quagmire.

- 我不確定下一步該怎麼走。
- I'm not sure where we should go from here.

- 或許我應該找其他人來做這筆生意。
- Maybe I should just take my business elsewhere.

- 除了你，我還是可以找其他人談。
- You're not the only game in town.

負面
- 這對我們兩個來說都是浪費時間。
- This has been a waste of my time and yours.

085 如何要求對方妥協

客氣

- 基於對彼此的尊重，相信我們可以討論出彼此都滿意的結果。
  - With our mutual respect, I'm sure we'll come to a decision that suits us both.

- 只要我們合作，一定可以達到預期的成果。
  - I know we can reach an agreement if we work together.

- 我們都想要最好的，那該如何做到呢？
  - We both want what's best—how can we make that happen?

- 妥協永遠是一個好辦法。
  - Compromise is always the way to go.

- 我們不都是在尋求雙贏的局面嗎？
  - Aren't we all looking for a win-win, here?

- 如果你同意，我們就這樣敲定吧！
  - If you agree, let's shake hands on it.

- 我們應該秉持退讓的精神，讓事情繼續走下去。
  - Let's move forward in the spirit of compromise.

- 如果能達成協議，對每個人都有好處。
  - Coming to an agreement is the best thing for all concerned.

- 我想我們彼此都需要做出些許讓步。
  - I think we both need to accommodate each other, here.

- 我真的希望能在這件事上達成共識。
  - I'd really like to come to an understanding about this issue.

- 做出對大家都好的決定，你不會後悔的。
  - You won't regret coming to a decision that benefits all parties.

- 我想之前的協商已經讓事情有了很大的進展，現在是做最後決定的時候了。
  - I think we've achieved a lot through our negotiations, but it's decision time.

- 只要你願意接納我的意見，我也願意接納你的。
- I'm willing to listen to you if you listen to me.

- 我們各自再做一些努力吧！
- Let's both make an effort.

- 我們應該做出一個決定，然後往下一個議題討論。
- I think we should agree to settle and move on.

- 希望你能至少把我的情況納入考慮中。
- I hope you're willing to at least consider my parameters.

- 天底下沒有完美的辦法，有得必有失。
- There's no magic solution; it's all give-and-take.

- 我們各自妥協吧！這是唯一的辦法。
- Meet me in the middle—it's the only answer.

- 只要你我各退一步，那就皆大歡喜了。
- We'll be fine if I give a little and you give a little.

- 我們各自捨棄一半的堅持，事情就成了。
- Let's each go halfway and we'll be done.

- 實在無法想像我們竟然沒辦法達成協議。
- I can't imagine us not coming to an agreement.

- 不談妥這件事情，回去我沒辦法對老闆／家人／同事交待啦！
- I can't go back to my [boss/family/colleagues] without settling this.

- 沒有所謂的完美方案，所以我們就這樣敲定，把事情解決了吧！
- There is no perfect solution, so let's just get it over with and settle.

- 我願意考慮交換一些條件，你呢？
- I'm willing to engage in some give-and-take—are you?

- 我們至少各退一步吧！
- At least meet me halfway.

- 最後你總是會有讓步的時候。
- I guess we'll get to that place when you're finally willing to compromise.

- 我們現在就把事情一次談定吧！
- Let's get this done once and for all.

- 與其這樣漫無止盡的討論下去，我們不如來丟銅板決定吧！
- Let's just flip a coin. It's better than going on and on about it forever.

- 拜託，這一次就配合我吧！
- Come on, work with me on this.

強硬

- 我們就打開天窗說亮話，省得浪費大家的時間。
- Let's cut to the chase and quit wasting everyone's time.

解決問題

善於找出問題、解決問題,將擬定的方案化
為行動……只要掌握幾個要訣,就能讓你從
容不迫的面對問題、解決問題,甚至還能夠
提供其他人方向,成為同儕間的佼佼者。

溝通哪些事

- 當需要處理／面對問題時
- 請他人幫忙時
- 簡化複雜的問題時
- 和遭逢困難的人對話
- 如何談論目前遇到的問題
- 如何談論一個過去的問題
- 準備提議一項計畫
- 當你需要大家冷靜時
- 如何警告他人
- 適當表達抱怨
- 如何回應他人的抱怨
- 當他人抱持負面態度時
- 當複雜事情過度簡化時

> 所謂領導者就是無論在先天才智
> 以及後天性格上都願意去解決問題的人。
> ——哈蘭克里夫蘭（美國外交官）

　　優秀的領導者善於解決問題，能夠化方案為行動，絕對能引起他人的注意，而且還能夠提供其他人方向。下面幾個要訣，能讓你從容不迫的面對並且解決問題：

1 分析判斷

　　盡可能發問，找出真正的問題是什麼。對策再怎麼高明，如果不對題也沒有用！就好像醫生診斷病人時，總會提出許多問題。透過發問，醫生才能獲得有用的資訊，進而做出診斷。別怕提出「笨問題」，盡可能蒐集資訊就對了。

2 宣導

　　一旦了解問題後，就必須開始幫助其他人了解問題嚴重的程度，還有不及時解決所可能造成的傷害。通常這包括了充分討論問題的各層面、設想任何可能的後續發展，還有說服大家一起參與解決問題。

❸ 擬訂策略

如果你做不好策略，那麼計畫注定會失敗。首先依次列出解決問題的流程步驟。也盡量不要下達命令，應該讓所有人都認為自己也是找出解決方案的人之一。如果有人反對，試著讓這些人了解你的解決方案對他們個人也有好處。

❹ 分派任務、追蹤進度

確認你的團隊掌握住整個問題的情況，而且能有效地將你的方案化為行動。別害怕分派任務給別人，你自己必須定期追蹤確保事情的進度。

086 當需要處理／面對問題

迫切

- 我們必須馬上找出解決的辦法，越快越好。
- We need to find a way out of this immediately, if not sooner.

- 現在，我們眼前有個一定要馬上解決的問題。
- We have a problem on our hands that must be solved, now.

- 我們現在就要想出辦法！
- We've got to figure this out right now.

- 我們必須盡快找出答案。
- We need to find an answer to this asap.

- 今天之內，我們一定要找出問題的根源。
- We must get to the bottom of this by the end of the day.

- 這種關鍵時刻，不達成協議絕不罷休。
- We can't go on without an agreement on this crucial point.

- 一定可以解決眼前的難題，我們只需要把辦法找出來。
- There must be a way out of this sticky situation. Let's find it.

- 我們一定要先全盤掌握眼前的問題，才能繼續往下討論。
- We can't move forward without coming to grips with the current problem.

- 要解決問題，就得馬上改變策略。
- To solve this problem, we need to immediately change tactics.

- 大家都同意我們必須找出一個快速有效的對策吧！
- I think we can all agree that we need a quick solution to this problem.

- 我們必須盡快把問題解決。
- We need to fix this as quickly as possible.

- 大家一定要團結合作，才能迅速解決問題。
- In order for the problem to be solved quickly, we must work together.

- 目前的首要任務是盡快找出問題的核心。
- Our priority is getting to the heart of the issue as soon as possible.

- 我們越能務實面對問題，就能解決越多問題。
- The more practical we are, the more ground we can cover.

- 讓我們先跳出框架，從各種角度來思考。
- Let's think outside the box for a moment.

- 看來目前的對策沒有任何效果，如果我們…
- Since what we've been doing isn't helping, what if we....

- 這個辦法行不通，我們需要試試別的辦法。
- This approach isn't working. We need to try something different.

- 我們必須對這個問題有更深入的了解。
- I think we need more clarity on the issue.

- 我們必須重新思考對策。
- We need to rethink that solution.

- 麻煩大了！我們現在該怎麼做呢？
- What a quandary! How do we get out of it?

- 現在唯一的辦法就是從更根本的層面來解決問題。
- The only way out is to deal with this issue in greater depth.

- 唯一的出路就是正面迎戰。
- There's only one way around this, and that's through it.

- 讓我們一起找出大家都能接受的解決辦法吧！。
- Let's settle on a solution that we can all be comfortable with.

- 我們應該合作找出答案，而不是各自為政。
- Rather than getting off track, we should work together to find a solution.

- 在採取動作之前，我們應該蒐集更多資訊。
- We need more facts about the situation before we do anything.

- 現在的進度躊躇不前，我們何不試試其他的方法？
- We're not getting much done this way. Why not try something else?

- 如果我們換個角度去想呢？
- What if we took this in a different direction?

- 應該是徹底了解問題的時候了，所以我建議…
- It's time to get better acquainted with this problem, so I suggest...

- 明明該…的時候，我們卻還在這裡浪費時間。
- We have been doing this when we should have been...

- 老是做一樣的事卻希望會有不同的結果，根本就是癡人說夢。
- Doing the same thing repeatedly and expecting different results is the definition of insanity.

- 如果我們不能解決問題，只會讓問題更加惡化。
- If we're not part of the solution, we're part of the problem.

- 眼前該做的是深入了解問題，而不是匆忙找到答案。
- The way forward is to dig deeper, not rush to a solution.

- 別這麼急，謹慎行事總比事後後悔來得好。
- Let's not rush this—better safe than sorry.

- 用急就章的辦法暫時掩蓋問題，反而是最糟糕的。
- The worst thing we could do is put a temporary bandaid on the issue.

- 我要全面性的對策，而不是暫時的修補。
- I want a comprehensive solution, not a temporary fix.

- 今天大家先好好休息，明天再繼續討論吧！
- Why don't we all sleep on it and reconvene tomorrow?

- 或許我們應該靜觀其變。
- Maybe we should let the problem solve itself.

放任

- 有時候不行動反而是最好的行動。
- Sometimes doing nothing is the best solution.

087 請他人幫忙時

專業

- 請問你願意幫忙我個忙嗎？
- Would you be so kind as to render your assistance on this matter?

- 很抱歉打擾你，但是我真的需要你的幫忙。
- I apologise for the interruption, but I really could use your help.

- 我自己真的無法完成這個案子，請問你可以幫點忙嗎？
- I can't see an end to this project— would you be willing to lend a hand?

- 你介意幫忙處理我手上的這一團亂嗎？
- Do you mind helping me sort out this mess?

- 請問你有空嗎？我迫切需要你的專業知識。
- I'm in dire need of your expertise—do you have a moment?

- 我不知道該怎麼解決這件事情？你有任何想法嗎？
- I don't yet know how I am going to fix this; do you have any ideas?

- 我需要找出一個解決辦法。你的想法是？
- I need to find a solution to this problem; what are your thoughts?

- 或許你就是那個能解決問題的人。
- You may hold the key to the answer.

- 請幫我一起找出一個正確的對策。
- Help me find the proper solution to the problem.

- 我已經無計可施了。如果你能幫忙，我會非常感謝。
- I'd appreciate it if you would help me out with this—I'm stumped!

- 我需要你的協助，否則不可能解決眼前的難題。
- I need your support or I'll never get out of this dilemma.

- 我似乎怎麼做都不對，或許你知道該怎麼辦？
- This seems to be slipping through my fingers—maybe you can do better?

- 不管你的意見是什麼，我都想聽聽。
- I'll listen to any input you might have.

- 我希望你能想出一個對策。
- I'd love it if you proposed a solution.

- 你願意幫忙嗎？拜託了。
- Would you help me out, please?

- 你可以在這件事情上助我一臂之力嗎？
- Can you give me some support on this?

- 你想到的辦法是什麼？
- What do you propose as a solution?

- 希望你能幫我度過這個難關。
- If only you could help me get out of this bind!

- 你心裡有什麼對策嗎？
- Do you have a solution to this mess?

- 我需要你的聰明才智／豐富經驗來幫我找出答案。
- I need your brilliant mind/vast experience to help me figure this out.

- 請幫幫我！
- Help me out, here!

直接
- 有興趣一起來嗎？
- Care to take a crack at it?

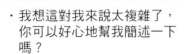

簡化複雜的問題時

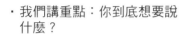

圓滑

- 我想這對我來說太複雜了，你可以好心地幫我簡述一下嗎？
- I'm afraid this is too complex for me—would you be so kind as to simplify the matter?

- 我們講重點：你到底想要說什麼？
- Let's keep this simple: what exactly do you mean?

- 你可不可以只列出真正重要的資訊？
- Is there any way you can outline just the salient facts?

- 我希望能夠理解你在說什麼，但我就是聽不懂。
- I wish I could grasp what you are saying, but I can't.

- 你複雜的理論對我來說實在太高深了點。
- I'm afraid your complex approach is far too sophisticated for me.

- 可以簡化你的想法嗎？我聽不到重點。
- Would you streamline your thoughts? I'm having trouble getting your point.

- 別把事情複雜化，我比較喜歡簡單明瞭。
- Please don't get too complicated. I'm better when things are stripped down.

- 我喜歡看事情的重點，你呢？
- I like to keep things simple—how about you?

- 麻煩說重點，謝謝。
- Just the main points, please

- 對話越簡潔有力越好。
- The more concise the dialogue, the better.

- 單純是目前最好的狀態。
- Simplicity is the best course for now.

- 讓我們回歸基本面，好嗎？
- Let's stick to the basics, okay?

- 我想這已經超過我所能思考的程度了。
- I'm afraid this is all over my head.

221

- 你為什麼現在才來講這些事情呢？
- Why do you want to go over all that right now?

- 我們應該著重在問題的基本架構。
- Let's stay within the basic framework of the issue.

- 如非必要，千萬不要把事情複雜化。
- Let's not complicate matters unnecessarily.

- 我們不要討論那些枝微末節的小地方了。
- Let's not go down useless tangents and dead-ends.

- 我們盡量簡單扼要，好嗎？
- Let's be more brief, shall we?

- 請注意基本要點。
- Please confine yourself to the fundamentals.

- 細節不重要，我們看大方向就好。
- The details aren't important—let's just go over the basics.

- 請用三句話告訴我發生了什麼事。
- Tell me exactly what's going on in 25 words or less.

- 告訴我問題的重點就好。
- Just give me the basic outline of the issue.

- 這真是太複雜了，你可以重說一次嗎？
- This is getting way too complex for me—can we start over?

- 如果我們只抓重點，問題就容易理解得多。
- This issue would be more easily understood if we kept things straightforward.

- 我只有幾分鐘，說重點。
- I only have a few minutes, so get to the point.

- 這位夫人，只要告訴我事實就好。（玩笑口吻）
- Just the facts, ma'am [joking]

- 別兜圈子說話，我沒有時間。
- Cut to the chase—I don't have a lot of time.

- 你這些枝微末節的小事快把我煩死了。
- You're killing me with all the excruciating minutiae.

- 記得，說話越短越淺顯越好。
- Remember, KISS means "keep it simple, stupid."

- 你可以說重點嗎？
- Can you just get on with it?

直接
- 省下那些沒用的廢話。
- Cut out the useless babble!

089 和遭逢困難的人對話

親密
- 在你這麼難過的時候，有什麼我可以幫忙的嗎？
- How can I best help you during this difficult time?

- 我真的很關心你，有任何需要幫忙的地方請讓我知道。
- I am really concerned about you and I want to help in whatever way I can.

- 你需要幫忙的時候，我會一直在你身邊。
- I am here for you if you need anything at all.

- 我有過類似的經驗，完全可以理解你的感受。
- I have been where you are now and I completely understand.

- 我完全了解你的痛苦。
- I hear you and feel your pain.

- 我知道你現在很容易感到絕望，但是我向你保證…
- While it is easy to feel helpless in times like these, let me assure you that...

- 我相信只要你願意討論，我們可以一起度過難關。
- I believe we can overcome anything by talking things through.

- 這件事遲早會過去的。
- This, too, shall pass.

- 接受事實吧！人生就是千瘡百孔。
- Let's face it, life is hard.

- 天總有不測風雲，你說是嗎？
- Life sure has a way of blindsiding you, doesn't it?

- 不管多麼渴望，我們還是無法逃離現實。
- We can't escape reality, as much as we want to.

- 雖然每個人都有一堆問題，但我們還是要繼續走下去。
- We all have to carry on despite personal problems.

- 在這種時刻，我們一定要保持堅強。
- In moments like these, we must be strong.

- 只要你能熬過去，就會變得更堅強。
- That which doesn't kill you makes you stronger.

- 無法改變的事我們只能接受。
- You just need to accept what you can't change.

- 每個人都會遇到困難，你也不例外。
- Everyone goes through tough times; you're no different.

- 每件事的發生都有原因。
- Well, they say that everything happens for a reason.

- 你會沒事的，我保證。
- Oh, you will be fine, I just know it.

- 你知道，每個人都有痛苦的時候。
- Nobody has a corner on suffering, you know.

疏離

- 挺起胸膛，事情會好轉的。
- Buck up—it'll get better.

090

如何談論目前遇到的問題

正面

- 我們討論一下該怎麼面對和解決這個問題吧！
- Let's go over what needs to be dealt with and fix it.

- 我相信討論可以幫助我們克服任何困難。
- I believe we can overcome anything by talking things through.

- 只要我們團結合作，一定可以克服難關的。
- We can overcome this challenge if we work together.

- 我們應該重新檢視一下所有的細節，才能更有效的解決問題。
- We should go over the details to deal with this situation better.

- 想要脫離眼前的困境，我們一定要深入討論。
- A discussion needs to happen in order for us to get out of this quagmire.

- 如果我們集思廣益，一定可以解決這個難題。
- We can solve this brain teaser if we put our heads together.

- 我們再來檢視一次目前遇到的困難。
- Let's take another look at the problem.

- 我們需要進一步的討論。你有幾分鐘嗎？
- We need to discuss things further. Do you have a minute?

- 我們是身陷難關沒有錯，目前最好的辦法就是討論可以怎麼做。
- Sure we're in a dilemma. Discussing our options will help.

- 或許多點討論可以幫我們釐清目前的狀況。
- Perhaps if we talk more we can sort it all out.

- 我們應該多多了解眼前遇到的難關。
- We should get better acquainted with this challenge.

- 現在有很多需要學習的新資訊，時間有限，我們就從最根本的著手吧！
- We have a lot to learn and not a lot of time. Let's get to the bottom of it.

- 這個問題讓大家都不好過，我們好好討論一下吧！
- What's happening isn't making anything easier. Let's discuss it.

- 這一次的討論應該會有所幫助吧！
- This discussion should help somehow.

- 如果我們動作不快一點，之後只會有更多的麻煩。
- If we don't work quickly, there will be a lot more trouble downstream.

- 不快點處理現在的狀況，就等著面對真正的問題吧！
- We've got a real problem if we can't clean this up soon.

- 再不快點解決這件事，後面會有更大的問題。
- If we don't solve this quickly it's going to lead to more serious trouble.

- 我們不能再讓情況繼續惡化下去。
- We can't afford to let the situation get any worse.

- 我們要阻止問題越演越烈。
- We can't let the situation boil over.

- 再不立即行動，問題只會繼續惡化。
- Without direct immediate action, the problem will only get worse.

- 問題不會自己解決的。
- This problem isn't going to solve itself.

- 我們離災難爆發就只差一步。
- We're on the verge of disaster, here!

- 都大禍臨頭了，卻沒有任何人在乎。
- The Titanic is sinking and nobody seems to care!

負面

091

如何談論一個過去的問題

正面

- 我們從那件事情上學習到許多，也成長了不少。
- We learned a lot from that and are stronger for it.

- 我們今天的成就來自於過去那段困頓的時光。
- Those hard times are what made us what we are today.

- 幸好，以前的那些難關都過去了。
- Thankfully, all of that drama is behind us now.

- 以前的困難沒有擊倒我們，而是讓我們變得更堅強。
- That which did not kill us just made us stronger.

- 那個難關如今看來已經是前塵往事了。
- That challenge already seems like it's part of the distant past.

- 我們遇到的那些問題都已經過去了。
- The troubles we faced are already in the rearview mirror.

- 還好，我們都走過來了。
- Fortunately, we're past all that now.

- 以前那些問題讓現在的我們更有智慧。
- The issues of the past only make us wiser in the present.

- 雖然面臨了各種挑戰，但我們相信最後一定都能克服。
- We faced difficult challenges, but we always knew we'd overcome them.

- 過去那些問題現在都已經是歷史了。
- The problems we encountered will be relegated to the history books.

- 從過去的難題中我們學到了許多，而現在最需要的則是往前看。
- We learned a great deal from that, but we need to look to the future.

- 我們不能改變過去，但還來得及創造更好的未來。
- We can't erase the past, but we can make a better future.

- 雖然那些困難都過去了，但對每個人都造成或多或少的傷害。
- While our troubles are behind us, no one escaped unscathed.

227

- 如果不能從過去中汲取教訓，那你註定會重蹈覆轍。
- If you don't learn from history you are doomed to repeat it.

- 很遺憾我們無法改變過去。
- Unfortunately we cannot rewrite the past.

- 發生過的事無法改變，我們只能繼續往前邁進。
- What's done is done; we just need to move on.

- 以前我們處理的不好，下次一定會進步。
- We didn't handle that well; we need to do better next time.

負面
- 有時我們似乎註定會一再犯下相同的錯誤。
- Sometimes it seems like we're doomed to repeat our past failures.

092 準備提議一項計畫

強勢
- 我相信我的計畫能帶來最大的贏面。
- I believe my plan gives us the best chance for success.

- 相信你會同意我的計畫是最好的選擇。
- I think you'll agree that my plan is the best choice.

- 我的提議是你唯一的選擇。
- The only option is to listen to my suggestion.

- 過去的經驗讓我相信我的計畫會帶領大家朝正確的方向前進。
- Experience tells me that my plan will lead us down the right path.

- 如果採用我的新主張，相信能夠有更豐碩的成果。
- If we look at my new idea, I think we can accomplish a lot more.

- 我們試試新的辦法吧！
- Let's try something new!

· 我們需要新方向；誰站在我這邊？	· We need a new direction; who's with me?
· 我想要介紹大家一個新想法。	· I would like to introduce an idea into the mix.
· 我想提出一個新的方向。	· There's a new direction I'd like to propose.
· 我們不如這樣做做看？	· How about we do it this way?
· 只有縝密的計畫才能夠帶來成功。	· Only thorough planning can help us succeed.
· 請考慮我的計畫，相信會有幫助的。	· Please consider using my plan—I think it will help.
· 我們已經考慮過很多選項，但我個人傾向於…	· We've pondered many options, but I'm in favor of...
· 這的確是個困難的決定，所以我…	· Our choices are never easy. That's why I'm in favor of...
· 除了…之外，我沒有其他的建議。	· I have no other choice than to suggest...
· 如果你願意，我們可以試試我的計畫。	· I guess we can try my plan if you want.
· 雖然不一定會成功，但是…	· This probably won't work, but...
· 情況或許已經無可救藥，但是…	· It's probably hopeless, but...

被動

093 當你需要大家冷靜時

正式	· 拜託大家，一起抱著寬恕的精神合作下去吧！	· Please, let's all work together in a spirit of tolerance.
	· 大家都冷靜點，我們才能繼續討論下去。	· We can stay focused if we all just calm down.
	· 我們必須表現出自己的專業素養。	· We have to demonstrate that we can keep to the straight and narrow.
	· 或許我們應該重新分組再進行討論。	· Maybe we should regroup and get our thoughts together.
	· 我們的重點應該放在眼前的任務。	· We must concentrate on the task at hand.
	· 我們一定要繼續照正常的速度往對的方向前進。	· We must keep the peace to keep moving in the right direction.
	· 重點是我們的工作，而不是彼此之間的差異。	· The emphasis must be on the work, not our differences.
	· 請大家保持冷靜。	· I'm asking everyone to remain calm.
	· 請大家先做個深呼吸。	· Everybody take a deep breath, please.
	· 請保持冷靜，我們一定要討論出答案。	· Please stay calm—we need to find solutions.
	· 我們沒時間做無謂的爭執了。	· This is no time to fly off the handle.
	· 這樣吵鬧不休，我實在沒辦法專心。	· I can't keep my concentration with all this static.
	· 這種吵吵鬧鬧的環境根本不能靜心思考。	· It's impossible to hear yourself think with all this commotion.
	· 我沒辦法在這種環境下工作。	· I can't work in all this upheaval.

・大家不要為這件事大動肝火。	・Let's not lose our heads over this.
・請大家安靜。	・Everyone keep the peace.
・各位，請肅靜！	・Settle down, everybody!
・請各位稍安勿躁！	・Be still, everyone!
・大家冷靜一點！	・Keep cool, everyone!
・不要這麼火爆！	・Don't get your knickers in a twist!

 隨性

・每個人都給我安靜／閉嘴。　・Everyone, just calm down/pipe down/shut up!

094 如何警告他人

 含蓄

・你確定你要這麼做嗎？	・Are you sure you want to do that?
・這可能不是最好的選擇。	・You probably don't want to do that.
・你或許應該認真考慮是不是要這麼做？	・Maybe you should rethink this/sleep on it.
・我只是想確定你知道自己在做什麼。	・I just want to be sure you know what you're getting yourself into.
・我知道你想這樣做，但是我請你重新考慮。	・I'm aware that you know what you're doing, but please rethink this.
・我真的不建議你去…	・I would not advise you to...
・我會建議你…	・I would suggest that you...
・你不覺得…是更合理的決定嗎？	・Wouldn't it be more rational to...

- 你不認為這可能會是個錯誤的決定嗎？
- Do you think it's possible that this is a bad idea?

- 只有傻子才會這麼做。
- Only fools rush in.

- 我求你不要這麼做。
- I'm begging you to please avoid this.

- 如果你執意如此，結果就是…
- Here's what you're in for if you go that route.

- 這麼做對你沒半點好處。
- This does not portend good things for you.

- 保證這對你有害無益。
- This certainly won't bode well for you.

- 最後不會有好下場的。
- This won't end well.

- 如果你這麼想，就準備好面對嚴重的後果吧！
- With that kind of mindset, prepare yourself for the worst.

- 如果事情失敗了，別來找我哭訴。
- Don't come crying to me if things don't work out.

- 到時候事情出了問題我可不負責。
- I'm not responsible if things blow up in your face.

- 我不希望以後再來責備你。
- I don't want to say, "I told you so."

- 這完全是個爛主意，別說我沒警告過你。
- Terrible idea—don't say I didn't warn you.

直接

095 適當表達抱怨

 禮貌

- 我是出自於關心才告訴你。
- If I did not care, I would not say anything about it.

- 我想讓你了解我的顧慮。
- I want to give you the opportunity to address my concerns.

- 你做得很好，不過還有可以更進步的地方。
- You're doing well, but here is where you can improve.

- 我通常不太抱怨的；你呢？目前還滿意嗎？
- I'm usually not one to complain—are you happy with things?

- 從我的角度來看，…
- Look at it from my point of view...

- 我不想得罪人，但是事情真的不太對。
- I don't want to be disrespectful, but this just isn't right.

- 我不是要小題大作，但我真的不同意這件事。
- I don't want to make a fuss, but this isn't working for me.

- 我通常不發牢騷，這次要破例了。
- I normally don't whine, but I have to make an exception.

- 雖然我一向隨和，但這真的太離譜了。
- I usually don't gripe about things, but this has gone too far.

- 你對於這些事情一點意見都沒有嗎？
- Don't you have a problem with what happened?

- 或許你可以接受，但我不行。
- You may be okay with this, but I'm not.

- 如果你在乎我的感受，就會改變的。
- If you cared about my feelings, you'd change this.

- 我對於你…的行為感到失望／我不贊成你…。
- I object to/am disappointed in the way you...

- 有件事情我對你很不滿意。
- I have a bone to pick with you.

233

・你不應該這樣對別人的。	・This isn't the right way to treat people.
・你這樣做是不對的。	・This will never do.
・這完全讓人無法接受。	・This is completely unacceptable.
・你徹底失去了我的尊重。	・You have completely lost my respect.
・我受夠了。	・I'm just about done here.
・你徹頭徹尾搞砸了。	・You screwed up royally.
・我再也不要替你做事／和你往來／和你生活在一起了。	・I no longer wish to do business with you/stay married to you/work for you.

粗魯

096 如何回應他人的抱怨

體貼

・你不開心，我心情也不好。	・If you're unhappy, I'm unhappy.
・如果我和你遇到相同情況，一定也很生氣。	・If I had to deal with what you are describing, I would be upset, too.
・我了解你的感受，也向你保證我會把事情處理到你滿意為止。	・I understand your concerns and I promise I will resolve things to your satisfaction.
・這的確是很糟糕，我保證會改善這個狀況。	・I agree that this was terrible—I promise I will make it better.
・我懂你的心情，也會盡我所能的解決問題。	・I understand why you would feel that way—I will do what I can to address it.
・我會負責到你滿意為止。	・I will not give up until you are completely satisfied.
・我想知道所有的情況才能馬上改善問題。	・I want to know everything so I can correct the problem immediately.

- 請詳細解釋目前的問題，我才能盡可能幫你。
- Help me understand the issue so I can give you my best help.

- 了解你的顧慮後，我就可以開始處理了。
- Now that I understand what's bothering you I can begin to address it.

- 我們會盡最大的努力在最短的時間內解決。
- We'll do everything to fix this as soon as possible.

- 真的很抱歉，我們已經在處理了。
- Sorry about all that—we're working on it.

- 請你諒解這一切真的是無心之過。
- Please understand, this was completely unintentional.

- 我向你保證這次絕對是例外，不是常態。
- Let me assure you that this is the exception, not the rule.

- 我了解你說的問題，但我們能做的有限。
- I understand your problem, but we can only do so much.

- 這不是我的錯，不過我還是盡我所能的改善這個狀況。
- This isn't my fault, but I'm doing my best to make things better.

- 請你先耐心等候，我正在請示主管。
- You'll have to be patient. I'm checking with my superiors.

- 目前我真的無計可施。
- There is nothing I can do for you right now.

- 你了解狀況以後就會明白有多困難了。
- It's very easy to assume when we don't know.

- 你用這種方式說話，我很難幫得上忙。
- I can't do much to help when somebody is talking this way.

- 別這麼生氣，我已經盡力在處理了。
- Don't be so unpleasant—I'm doing the best I can.

- 你應該知道，抱怨是完全沒有用的。
- Complaining isn't very constructive, you know.

- 你的抱怨並不是當務之急。
- Your complaint isn't top priority right now.

- 破口大罵對事情一點幫助也沒有。
- Shouting/your attitude isn't making anything better.

- 沒有人是完美的。
- Nobody's perfect.

- 這不關我的事。
- Not my problem.

冷漠
- 我無所謂。
- I don't care.

097 當他人抱持負面態度時

同情
- 我完全理解你的心情，你的確有生氣的理由。
- I totally understand what you're saying and agree that you have a right to be upset.

- 我知道你今天很不順，有什麼我能做的嗎？
- I'm sorry you're having a bad day—what can I do to help?

- 雖然不容易，但我們試著保持比較正面的態度吧！
- I know it's tough, but let's try to stay positive.

- 從今天開始，我們試著把負面情緒轉為正面能量吧！
- Let's start today by turning our negatives into positives.

- 我們盡量維持建設性的想法和行為，好嗎？
- Let's focus on being constructive, okay?

- 我知道你今天很不好受，但是…
- I know you're having a rough day, but...

- 其實，不管什麼問題總是有解決的辦法。
- Every problem has a solution, you know.

- 所以你現在有什麼打算？下一步怎麼做
- So what are you going to do about it? What's your plan?

- 我知道這是一個難題，但你要想辦法克服。
- Sure there's a problem, but you need to push through it.

- 記住，正面的態度會帶來高昂的鬥志。
- Remember that your attitude has an effect on morale.

- 這種喪氣的話只會讓事情變得比原來更糟。
- This kind of talk is making things worse than they really are.

- 我們何必討論這種事呢？
- Why are we even talking about that?

- 你不覺得自己有點小題大作了嗎？
- Don't you think you're exaggerating a little?

- 這是你的感覺，但不是事實。
- Feelings aren't facts.

- 難道你一整天都要這樣垂頭喪氣的嗎？
- Are you going to be in this kind of a mood all day?

- 拜託，你這樣很不專業／很難相處／根本幫不上忙。
- Come on, you're not being very professional/helpful/easy to live with.

- 這件事已經讓你變了個人。
- You're letting this get the best of you.

- 你的憂鬱已經影響了我們大家的情緒。
- We're catching your negative vibe like a virus.

- 抱怨是做不成任何事的。
- Whining never accomplished anything.

- 現在不是你自憐自艾的時候。
- This is no time to be pessimistic.

- 不要老是負面思考。
- Don't embrace the negative.

- 你這樣只是自掘墳墓。
- You're just being a defeatist.

- 很多人都遇過更糟的狀況。
- A lot of people have it much worse.

- 你根本是在小題大作。
- You're making a big deal out of nothing.

- 現在不是說這些話的時間和場合。
- This isn't the time or the place for that kind of chatter.

- 你影響了整個團隊的士氣。
- You're harshing everyone's mellow.

- 你正在消耗這裡的正面能量。
- You're bringing down the energy of this place.

- 拜託，請不要把你私人的事帶進來這裡。
- Please, don't add your personal problems into the mix.

- 你的尖酸刻薄／負面思考讓我很生氣／沮喪。
- Your cynicism/negativity is making me depressed/tense/upset.

- 我現在真的沒心情聽這些。
- I'm really not in the mood for this right now.

- 我不想再聽下去了。
- I've heard just about enough of that from you.

- 我們該辦個派對慶祝你有多悲慘嗎？〈諷刺語氣〉
- Should we all throw a pity party for you? [sarcasm]

- 你是林黛玉再世嗎？〈諷刺語氣〉
- Should we call you Eeyore? [sarcasm]

- 人生就是這樣，接受現實吧！
- That's life—get over it.

- 我光聽你說這些就已經筋疲力盡了。
- I feel exhausted just listening to you.

冷漠
- 請你住嘴。
- Please just put a lid on it.

當複雜事情過度簡化時

體貼

- 這件事可能比你說的還要複雜些。
- This might be a bit more complex than the way you're describing it.

- 我想這件事應該還有另外一面吧！
- I think there's more than one side to this issue.

- 我知道簡單扼要的重要性，但是…
- I understand the advantage of boiling things down to their essence; however...

- 事實上這件事非常複雜，你不認為嗎？
- In reality, the issue is a lot more complex, don't you think?

- 這件事情或許需要更深入的調查。
- A more sophisticated inquiry into the issue might be more helpful.

- 細節才是主角。
- God is in the details.

- 這樣簡化一個複雜的問題不見得是好事。
- Cutting a complex issue down to its component parts is not always wise.

- 你根本連重點都還沒講到就結束。
- You cut to the chase almost before the movie began.

- 你不能過度簡化這些事。
- You can't simplify things to such an extreme.

- 這是整件事的簡陋版本。
- That's basically the stripped down version of the issue.

- 整件事情已經被你刪得面目全非。
- You can't prune everything off and still expect it to be a tree.

- 我覺得你沒抓到事情的重點。
- I think you're dumbing this down.

直接

- 你看這件事的眼光很短淺。
- That's a myopic way of looking at it.

表現風度禮貌

想要成為一個有風範的人，時時保持風度禮貌是絕對必要的，把握住幾個訣竅，記得己所不欲、勿施於人，常常保持微笑，對所有人一視同仁，還要善用肢體動作、聲音語調，不隨他人八卦，就可以讓你隨時顯得風度翩翩。

溝通哪些事

- 如何請別人幫忙
- 如何幫忙別人
- 當有人請你幫忙時
- 如何感謝他人
- 當有人感謝你時
- 如何誇獎某人
- 當他人誇獎你時
- 當有人邀請你時
- 如何為自己說過的話道歉

- 如何為你做過的事情道歉
- 如何為了忘記某事而道歉
- 如何為了遲到道歉
- 如何為了你的組員的錯道歉

> 對紳士而言，風度和勇氣缺一不可。
> ——西奧多‧羅斯福（Theoroe Roosevelt, 第二十六任美國總統）

就算所有事情都順利運作，領導人仍然必須在日常生活的各個層面表現出自己的特質魅力，包括風度禮貌。但往往一不注意就忘了保持風度禮貌。這裡有幾個訣竅讓你隨時顯得風度翩翩：

❶ 己所不欲，勿施於人

相信你一定看過辦公室裡的同事怒氣沖沖地對另一名同事大聲咆哮。其實只要一點點的同理心與尊重，甚至只要多一點禮貌，情況就會大為好轉。如果你發現自己處在失控的邊緣，先深呼吸，然後離開現場冷靜一下，等情緒恢復以後再回來繼續討論。

❷ 適時微笑

簡單的微笑能讓你成為一個親切好相處的人。不管何時，一個微笑通常也能帶動他人的笑容，自然而然散播歡樂的氣氛。真正頂尖的領導人絕對不會顯得高高在上或是令人心生畏懼的。

❸ 一視同仁

在談話中展現良好風度的人多半對所有人一視同仁，能夠接納不同

的觀點和表達方式。真正的領導人也會避免任何帶有歧視或者性色彩的言語。你的語言不僅能改變自己的生活，甚至影響到周遭所有人的生活。

④ 善用肢體動作、聲音語調

善於溝通的人一定懂得運用肢體語言。約翰柏格分析過，人與人之間的溝通有 93％ 屬於肢體和表情聲音，只有 7％ 是文字內容。正因如此，肢體語言其實最能傳達一個人真正的心情與想法。你可以在鏡子前檢視自己的肢體動作，最好是能錄成影片來看（警告：你看了以後可能會大吃一驚）。切記，和人說話時眼睛要看著對方，同時肢體動作不要太過浮誇。音量適中就好，並且避免情緒化的表達方式，那會模糊了溝通重點並降低你的可信度。

⑤ 不八卦

雖然在茶水間或走廊上分享同事們的八卦總是令人興味盎然，但如果話傳回當事人耳裡，麻煩可就大了。你可能覺得和當時一起八卦的同事們培養了情誼，但你愛八卦的習慣會讓周圍的人感到不舒服。真正有風度的人不會針對別人說長道短。拒絕八卦也會帶來其他人的信任：他們知道你不會在背後討論任何人，包括他們自己的八卦。

⑥ 不打斷他人

如果你習慣打斷別人的談話，別人會認為你是個粗魯、好辯、自大的人。真正有風度的人會先讓其他人闡述完自己的觀點。如果你對他人的發言有些問題想請問，先寫下來，等到適當的時候再提出。

099 如何請別人幫忙

正式

你能幫我一個忙嗎？	Would you be so kind as to render your assistance in this matter?
只要你願意幫忙，我就很感激了。	I would be grateful for any help you could offer.
你的協助對我意義非凡。	I would hold your offer of help in the highest regard.
我真的不想打擾你，但能否請你…？	Far be it from me to bother you, but would you...?
這只是我單方面的請求，你不答應也沒關係。是否能請你…？	You're certainly under no obligation, but would you...?
如果你願意的話，是不是可以請你…？	Only when you get a chance, would you...?
你有空的話，能否請你…？	When you have a moment, would you...?
如果不是太麻煩的話，能不能請你…？	If it doesn't take too much time, would you please...?
不曉得你能不能…？	I was wondering if you would possibly...?
如果你能幫忙，我真的感激不盡。	Helping me right now would be the ultimate act of kindness.
請協助我一起解決現在面臨的問題。	Please help me get through this minor catastrophe.
就算是幫一點小忙，我也很感謝。	I would appreciate a little assistance with this.

- 你願意幫我個忙嗎？應該花不了太多時間。
  - Would you mind helping me with this? It shouldn't take long.

- 如果你不介意，可以請你…？
  - If you're okay with it, would you please...?

- 你願意助我一臂之力嗎？
  - Would you please lend me a hand?

- 拜託！我需要你幫忙。
  - I need a favor, please!

- 幫我個忙，好嗎？
  - Do me a solid, won't you?

- 你先幫我這個忙，我以後一定還你的。
  - I'll owe you one for sure.

- 如果是我的話，這個忙一定會幫的。（玩笑口吻）
  - You know I'd do it for you! [joking]

 隨性

- 救命啊！（玩笑口吻）
  - Man overboard! [joking]

100 如何幫忙別人

 正式

- 我永遠等著你開口。
  - I am, as always, at your disposal.

- 請讓我盡一份心力。
  - Please allow me to be of assistance.

- 只要你開口我一定會盡力。
  - Ask and you shall receive.

- 如果需要我幫忙，請隨時開口。
  - If you should ever need my help, please feel free to let me know.

- 你希望我如何幫你？
  - How can I best help you?

- 我會盡全力協助你。
  - I'm here to support you in any way I can.

· 你放心，有我在。

· You can always count on me.

· 如果我不幫你，還能算是朋友嗎？

· What kind of friend would I be if I didn't help you?

· 我知道你也會這樣幫我的。

· I know you would do it for me.

· 有任何需要幫忙的地方嗎？

· Is there anything I can do?

· 如果有我可以幫忙的地方，儘管開口。

· Let me know if I can do anything, okay?

· 你需要幫忙嗎？

· Do you need a hand with that?

· 你看起來需要一點協助，是吧？

· You look like you could use some help—am I right?

· 幫忙絕對沒問題，帳單隨後寄到。（玩笑口吻）

· I can help you but it's going to cost you. [joking]

· 如果不幫你，我可是會良心不安的。（玩笑口吻）

· If I didn't help you, how could I sleep at night? [joking]

· 需要幫忙就跟我說，好嗎？

· Give a holler if you need anything, okay?

隨性

當有人請你幫忙時

同意		
	・我很樂意幫這個忙。	・I would be delighted to assist you in this matter.
	・當然，我很樂意。	・Of course, I'd love to!
	・我正等你開口呢！	・I was hoping you'd ask.
	・我絕對會幫忙的。	・I'd be glad to help.
	・你放心，有我在。	・You can always count on me.
	・你只要說一聲就好了。	・There's no need to ask me twice.
	・你也總是在一旁支持著我，不是嗎？	・You're always there for me, aren't you?
	・我想我能幫得上忙。	・I guess I can try to help.
	・好，但我怕自己幫不上太多的忙。	・Okay, but I probably won't be of much help to you.
	・我盡量，但還是別抱太大的期望。	・I'll see what I can do, but don't get your hopes up.
	・好，但是你欠我一次喔！	・Okay, but you really owe me one.
	・很抱歉，這件事我真的幫不上忙。	・I regret that I can't be of assistance in this matter.
	・雖然我很想幫忙，但真的沒辦法。	・I'd love to help you but I just can't right now.
	・我很想幫忙，但真的是分身乏術。	・I'd like to but I'm already up to my neck.
	・抱歉，我沒有時間。	・I don't have the time, unfortunately.
	・我不得不拒絕。	・I'm going to have to decline.

- 我想我沒辦法這麼做。 · I don't think that's something that I'm going to do.

- 為什麼我要幫你？ · Why should I?

- 你必須自己處理。 · You need to figure it out yourself.

- 你知道幫得越多越沒用是什麼意思嗎？ · Have you ever heard of *learned helplessness?

拒絕 · 給我滾開！ · Get lost!

* 編按：learned helplessness 是心理學術語「習得性無助」，指人或動物因不斷受挫，而對一切感到無能為力，陷入無助的心理狀態。

102 如何感謝他人

專業 · 我真的非常感激你所做的一切。 · I very much appreciate everything you've done.

- 對於你的付出，我十分感激。 · I greatly appreciate all your efforts.

- 非常感謝你。 · Thank you so much.

- 我永遠都欠你一個人情。 · I am eternally in your debt.

- 再怎麼跟你說謝謝也不夠。 · There's no way to thank you enough.

- 真的不知道要怎麼表達我的感謝之意。 · How can I ever express my gratitude?

- 我對你的感謝，真的是筆墨難以形容。 · Words cannot describe how grateful I am.

- 你所做的一切，我都會銘記在心。 · I'm so grateful for all you've done.

- 我該怎麼謝謝你的付出呢？
- How can I thank you for all your hard work?

- 像你這樣的人現在已經不多了。
- People like you are a rarity these days.

- 我想沒有人可以像你這麼幫忙的。
- I don't think anyone else could have helped as much as you.

- 我真的打從心底感激你。
- I would like to thank you from the bottom of my heart.

- 要不是你幫忙，我到現在都沒辦法喘口氣呢！
- Without your help I would have been floundering.

- 我真的很感謝你為我做的一切。
- I really appreciate all you've done for me.

- 你真是個好人。
- That was very kind of you.

- 這次是我欠你。
- I owe you one.

- 你真是太好了，謝謝。
- You're a gem, thanks!

隨性
- 你好讚！
- You rock!

103 當有人感謝你時

正式

・你真的不用這麼客氣！	・You are most welcome.
・我很高興能幫得上忙。	・I'm glad I could help.
・至少這是我能做的。	・It was the very least I could do.
・我很開心自己能盡一份力。	・I am so glad I could be of service.
・我隨時都能幫忙的。	・I would do it again in a heartbeat.
・你開心，我也開心。	・If you're happy, I'm happy.
・我隨時都願意幫忙。	・It's always a pleasure.
・這是我的榮幸。	・My pleasure.
・沒關係。	・No problem.
・我很樂意幫忙。	・Happy to help.
・不用客氣。	・You bet!
・拜託，不值得一提啦！（玩笑口吻）	・Pray, don't mention it. [joking]
・小事一樁啦！（玩笑口吻）	・Ah, it was nothing. [joking]
・下一次我就要收費囉！（玩笑口吻）	・Next time, I'll charge you! [joking]

隨性

如何誇獎某人

正式

- 我非常欣賞你。
- I admire you greatly.

- 我對你有很高的評價。
- I think so highly of you.

- 和你一樣的人簡直鳳毛麟角。
- People like you are a rarity.

- 你是我效法的對象。
- You're someone I would like to emulate.

- 一看就知道你是個有能力／很聰明／很有經驗的人。
- You're obviously a person of great skill/intelligence/experience.

- 你總是追求完美。
- You always strive for excellence.

- 你真的令人印象深刻。
- You are so impressive.

- 我非常佩服你。
- I admire you so much.

- 我打從心裡敬佩你。
- I take my hat off to you.

- 你是個真正會做事的人。
- You're a real mover and a shaker.

- 你是個不可多得的人才。
- You're one of a kind.

- 我覺得你很厲害。
- I think you're awesome.

- 幾乎沒有什麼事情可以難倒你吧！
- You're a winner.

- 你總是可以主動出擊，手到擒來。
- You're a real go-getter.

- 你是最優秀的！
- You're the best.

- 你是我心目中的英雄。
- You are my hero.

- 如果我可以選擇的話，我想變成你。（玩笑口吻）
- If I could be anyone else, it would be you. [joking]

表現風度禮貌 7

251

・如果能變成你，就算一天也好。（玩笑口吻）	・I wish I could be you for a day. [joking]
・你很優秀！	・You're great.
・你很棒！	・You're awesome.
輕鬆 ・你太神了吧！	・You rock!

105 當他人誇獎你時

接受 ・我本來就是這樣啊！	・Tell me something I didn't already know.
・是啊！我真的是很厲害，對吧？	・Yes, I am fabulous, aren't I?
・你這麼說真的是太客氣了。	・How kind of you to say so.
・謝謝，你真是好心。	・Thank you, that's very nice of you to say.
・謝謝你的誇獎。	・Thank you for the compliment.
・這些話從你嘴巴說出來，對我來說是最大的讚美。	・That is praise indeed, coming from you.
・別再說了，我的尾巴都忍不住翹起來了。（玩笑口吻）	・Oh stop, before my head gets too big. [joking]
・你真是我的伯樂啊！（玩笑口吻）	・Ah, it takes one to know one. [joking]
・很感謝你這麼說。	・I truly appreciate that.
・我真的不知道該說什麼。	・I don't know what to say.

・拍馬屁是沒有用的。（玩笑口吻）
・Flattery will get you nowhere. [joking]

・拜託，你說得我都不好意思起來了。
・Stop, you're making me blush.

・這真的不是我一個人的功勞。
・I can't take all the credit, you know.

・其實我只是得到了許多幫助而已。
・It wasn't just me—I had a great deal of help.

・團隊裡是沒有個人的。
・There's no "I" in "team."

・我不敢居功，但還是謝謝你。
・I can't take credit for that, but thank you.

・我真的不敢當。
・I really can't take credit for it.

・就算沒有人知道，我還是會做一樣的事。
・I'd do the same thing even if no one were around to notice.

・我只是做了每個人都會做的事罷了。
・I only did what any ordinary person would do.

・你的讚美我真的愧不敢當。
・I don't deserve such flattery.

・這真的沒什麼特別／大不了的。
・Oh, it was nothing special/no big deal.

推辭　・拜託，事情根本不是這樣的。
・C'mon, that's not true.

表現風度禮貌

7

106 當有人邀請你時

接受

・謝謝你邀請我。	・Thank you so much for thinking of me.
・能夠參加是我的榮幸。	・I would be honored to attend.
・我很樂意參加。	・I'd love to come.
・我一定會到。	・You can count on me to be there.
・我很期待。	・There's nowhere else I'd rather be.
・謝謝你邀請我，我很樂意參加。	・Thank you so much for the invitation—I'd be delighted.
・我會準時到場的。	・I will be there with bells on.
・我一定會出席。	・Consider this my R.S.V.P.!
・我現在就能回覆你我一定會到。	・You don't have to ask me twice.
・我已經把那個時段保留給你了。	・I've already got it marked down in my book/calendar.
・我很樂意參加，你已經有賓客名單／活動流程了嗎？	・I'd like to be there—is the agenda guest list set yet?
・我會參加！可以請你寄給我活動細節嗎？	・I'd like to attend. Would you e-mail me the details?
・我必須更了解活動內容才能確定是否能出席。	・I can only accept if I know a little more about the event.
・我很想參加，但得先查查我的行事曆。	・I'd like to but I need to check my calendar.
・先讓我確認一下，我怕當天已經有了其他安排。	・I really don't want to double book, so let me check.

- 最近有很多事在忙。我可以明天回覆你嗎？
  - I have a lot going on right now. Can I get back to you tomorrow?

- 我這幾天非常忙，不確定到時候能不能參加。
  - I'm so busy these days—not sure if I can make it.

- 如果能改天的話，或許我可以參加。
  - If it was on another day, maybe.

- 應該是沒辦法，我確認後再跟你說。
  - Probably not—I'll have to get back to you.

- 抱歉我真的沒有時間，一點空檔都沒有。
  - Unfortunately I can't break free, not even for a moment.

- 很可惜我已經另有安排了。
  - Unfortunately I have previous engagement.

- 我不想缺席，但我這陣子真的是分身乏術。
  - I hate to turn you down, but I'm tapped out these days.

- 我很想參加，但當天剛好有其他事情。
  - I'd like to, but I have something else going on.

- 我必須對你說抱歉，我相信活動會非常圓滿成功。
  - I have to give you my regrets—but I'm sure you'll have a great time without me!

- 抱歉，我那天沒空。
  - I'm tied up that day, sorry.

- 我這陣子真的沒時間參加任何社交活動。
  - I just don't have time for a social life these days.

拒絕

107 如何為自己說過的話道歉

謙卑

- 我剛才說的話確實不對，請原諒我。
  - There's no excuse for what I said; please forgive me.

- 我真的很容易說錯話，非常抱歉。
  - I really have knack for saying the wrong thing. I am so sorry.

- 請接受我的道歉，剛剛我說了不該說的話。
  - Please let me apologize; that was an awful thing to say.

- 我老是說些不得體的話，請你原諒我。
  - I always put my foot in my mouth, it seems; please forgive me.

- 我剛才真是昏了頭才會那麼說，我很抱歉。
  - Clearly, I wasn't thinking clearly when I said that. I apologize.

- 對於剛剛說的話我很愧疚，請給我一個彌補的機會。
  - I feel horrible about what I said; let me make it up to you.

- 我希望剛剛沒有說那些話，真是糟透了。
  - I wish I could take it all back; I feel terrible.

- 我剛剛說的話完全沒經過大腦，可以給我機會重說一次嗎？
  - I spoke without thinking—can I have a do-over?

- 我是個笨蛋，才會說這些話。
  - I feel like a complete klutz about what I said.

- 我承認，剛剛說那些話是我不對。
  - What I said was wrong, I admit it.

- 我知道剛剛我說那些話是太過分了。
  - I recognize that I stepped out of bounds with that remark.

- 可以給我一個改過自新的機會嗎？
  - Please allow me to rectify the situation.

- 抱歉，我可以重說一次嗎？
  - I'd like to restate that, please.

・我可以收回剛剛說的話嗎？	・May I retract my statement?
・我真不敢相信自己會這麼說。	・I can't believe I said that.
・我真的不是故意要傷害你的。	・I didn't mean any harm, you know.
・可以讓我重說一次／收回剛剛的話嗎？	・Let me rephrase that/take it back.
・如果我的話沒有分寸，真的很抱歉。	・If I crossed the line, I apologize
・我剛剛的確說了令人無法原諒的話。	・There's no excuse for what was said.
・抱歉，我不知道你會這麼在意這些話。	・I didn't think you were going to take it so personally, sorry.
・抱歉，我今天的心情很糟。	・I was just having a bad day.
・我一時嘴快，抱歉。	・My tongue got the better of me.
・請不要把我剛剛說的話放在心上。	・Please pay no mind to what I said.
・我剛剛真的就這樣把話脫口而出嗎？	・Whoa—did I say that out loud?
・我想剛剛只是一時衝動。	・I guess I had a lapse in judgment.
・我想剛剛講這些話的時候，大腦正在罷工吧！（玩笑口吻）	・I think I had my brain in neutral when I said that. [joking]
・如果我說錯話會傷害了你，我道歉。	・I apologise if I misspoke or hurt your feelings in any way.
・如果我說錯了，請接受我的道歉。	・Accept my apologies if what I said was wrong.

- · 我真的是一時不小心才會這麼說。
- · I said what I said because you caught me off guard.

- · 我知道我錯了，但只要是人都會犯錯的嘛！
- · I know I made a mistake, but I'm only human.

- · 你從來沒有說錯話過嗎？
- · Have you never made a mistake?

傲慢
- · 別這麼敏感嘛！
- · Don't be so sensitive.

108 如何為你做過的事情道歉

謙卑
- · 請你原諒我，雖然我做的事情真的不可饒恕。
- · Please forgive me; there is no excuse for my actions.

- · 我的行為的確不可寬恕，請你原諒。
- · My actions were inexcusable; please forgive me.

- · 我確實把事情搞砸了，希望你能原諒我。
- · I really made a mess of things; I hope you can forgive me.

- · 我很抱歉那樣對你，請給我一個彌補的機會。
- · I am sorry I did you wrong—please allow me to make it up to you.

- · 我很後悔自己曾經做過這些事情，我保證絕對不會再犯了。
- · I regret it and I promise it will never happen again.

- · 我不知道該如何表達有多麼後悔過去的行為。
- · I can't tell you how sorry I am for what I did.

- · 希望你能原諒我以前的蠻橫無理。
- · I hope you can forgive me for my thoughtlessness.

- · 無法用言語表達我對做過的事的懊悔。
- · Words can't express how much I regret my actions.

- 我從來沒有想過要傷害／侮辱你。
  - I never, ever meant to hurt/disrespect you.

- 我絕對沒有傷害你的念頭。
  - Hurting you was the last thing on my mind.

- 雖然我非常抱歉，但傷害已經造成了。
  - Even though I'm sorry, I know that won't make it go away.

- 過去我不了解自己的錯誤，但現在我非常後悔。
  - Obviously I regret what I did, even if I didn't know it was wrong.

- 當時的我完全不知道自己在做什麼。
  - I guess I just wasn't thinking.

- 我會盡我所能不再犯同樣的錯。
  - I'll try to never do that again.

- 我非常懊惱，但不管如何，已經做過的事情是無法改變的。
  - I feel badly, but I can't undo what I did, unfortunately.

- 如果有機會，很多事我會用不同的方式處理。
  - There are a lot of things I would do differently if I could.

- 如果我過去曾經冒犯／傷害了你，我道歉。
  - If I stepped on your toes/hurt you/offended you, I apologize.

- 如果傷到了你，我很抱歉。
  - I'm sorry if I hurt you.

- 聽著，我只不過是個普通人。
  - Listen, I'm only human.

- 我也很不好受，難道你一點錯都沒犯過嗎？
  - I feel wretched, but haven't you ever made a mistake?

- 我又不是故意的。
  - I didn't do it on purpose!

傲慢
- 全都是我的錯！（諷刺口吻）
  - Mea culpa. [sarcasm]

109 如何為了忘記某事而道歉

謙卑

- 是我的不對，在此誠心跟你道歉。
  - It was inexcusable of me to forget; I humbly apologize.

- 我真的不是個健忘的人；我保證以後不會再發生一樣的事了。
  - It is so unlike me to be this forgetful; it will never happen again.

- 我希望可以回到過去再重來一次。
  - If I could turn back the hands of time, I would.

- 不論說什麼話都無法表達我的歉意。
  - Words cannot express how badly I feel.

- 我竟然忘記了，真的是太對不起大家了。
  - I'm so thoughtless to have forgotten.

- 你能夠原諒我這麼粗心大意嗎？
  - Can you ever forgive me for being so distracted?

- 我通常是很有條理的，真不知道這是怎麼發生的。
  - I'm usually very organized; I don't know how this happened.

- 我真不敢相信竟然會忘了這件事。
  - I just can't believe this slipped my mind.

- 我不知道自己到底在想什麼。
  - I don't know what I was thinking.

- 我很抱歉，之前真的是頭腦不清了。
  - I apologize; I'm just a lame brain.

- 你知道的，我有時候腦子真的不太靈光。
  - You know me—I'm just a flake.

- 請原諒我，我真的是忘記了。
  - Please forgive me—I simply didn't remember.

- 怎麼辦？我真的控制不了自己的健忘。
  - I can't help it if I'm so forgetful.

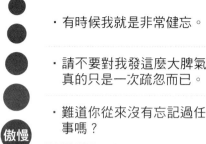

・有時候我就是非常健忘。

・I just get a little absentminded from time to time.

・請不要對我發這麼大脾氣，真的只是一次疏忽而已。

・Don't get on my case; it was an oversight and nothing more.

・難道你從來沒有忘記過任何事嗎？

・Have you never forgotten anything in your life?

傲慢

110 如何為了遲到道歉

謙卑

- 很抱歉我遲到了，我以後絕對不會再犯！
  - I'm so sorry for being late; I won't let it happen again!

- 謝謝你的等候，很抱歉我遲到了。
  - Thank you for waiting; I apologize for my tardiness.

- 我真的不應該讓你等我，請原諒我。
  - I have no valid reason for keeping you waiting; please forgive me.

- 我的遲到真的不可原諒。
  - There is absolutely no justification for my tardiness.

- 我錯了，我應該提早做好規劃的。
  - I have no excuse; I should have planned ahead.

- 我一向很準時，這次不知道是怎麼了。
  - I always strive to be punctual; I don't know what happened.

- 這真的不像我了，我從來不遲到的。
  - This isn't like me; I'm never late.

- 雖然我遲到了，但我會盡量彌補你。
  - Yes I was late, but I'll make it up to you

- 我想我該買支手錶了。（玩笑口吻）
  - I guess it's time for me to buy a watch! [joking]

- 謝謝你的耐心等候。
  - Thanks for your patience.

- 如果你再也不想看到我，完全可以了解。（玩笑口吻）
  - I understand if you never want to meet me again! (joking)

- 我希望你沒有在這裡站太久。
  - I hope you weren't standing here too long

- 我之所以遲到是因為最近生活真的一團亂，相信你可以體諒。
- I'm late because my life is so hectic; I'm sure you understand.

- 我知道這個理由很糟糕，但是我最近真的有太多事情要處理了。
- I know it's a lame excuse, but things have been hectic lately.

- 很抱歉我遲到了：〔理由〕。
- Excuse my lateness: [insert random excuse here].

- 你知道我這個人總是會遲到個十分鐘。
- You know me: I'm always 10 minutes late for everything.

- 準時真的不是我的優點！
- Punctuality was never my forte.

- 我好像老是遲到。
- It seems I'm always falling behind

- 我是遲到了，那又怎樣？
- I know I'm late, but so what?

- 難道你這輩子沒遲到過嗎？
- Have you never been late in your life?

傲慢

- 你能夠看到我已經是祖上積德了。（諷刺口吻）
- It's a wonder that you'll ever meet with me again. [sarcasm]

111 如何為了你的組員的錯道歉

謙卑

- 對最近發生的事，我個人會負起全部責任。
- I am personally responsible for what transpired.

- 這是我的錯！以後我會小心不再發生同樣的事。
- It was my fault; it will never happen on my watch again.

- 我的團隊的確有疏失，對此我向大家致歉。
- I know that we dropped the ball, and for that I apologise.

- 對於所發生的事致上最高歉意；我的團隊會盡全力補救損失。
- I am tremendously sorry for what occurred; this team will make up for it in full.

- 我很抱歉造成大家的困擾，請問我們該怎麼彌補呢？
- I apologize for the inconvenience; what can we do to make up for it?

- 如果有任何我們可以彌補的地方，我們團隊一定做到好為止。
- If there's anything we can do, this team is equipped to fix this.

- 對於這件事我們很內疚，也對所造成的傷害感到非常抱歉。
- We are disturbed this happened and are deeply sorry for any damages.

- 我們保證，像這樣的疏失絕對不會再度發生。
- We fully guarantee that this mishap will never repeat itself.

- 我了解事情的嚴重性，目前整個團隊正在盡力補救。
- I'm fully aware of the situation and we're working on it now.

- 以後我們絕對會更加小心謹慎的。
- We will definitely be more careful in the future.

- 請放心，我們已經從這次的錯誤中學到了教訓。
- Please rest assured that we have learned from our mistake.

- 目前我們正在盡一切的可能彌補錯誤。
- Please know we are doing everything we can to fix this.

・ 對於過去我們曾犯的錯誤，僅代表我的團隊致上深深的歉意。

・ I apologize on behalf of the team for any errors that might have been made.

・ 我承認我們的確有很大的改善空間。

・ I admit that we've been far from perfect.

・ 我的團隊一定會負起該負的責任。

・ My group must take responsibility for its actions.

・ 身為團隊的一份子，我們都對所犯的錯誤感到內疚並會負起全部責任。

・ As a group, we are ashamed of what occurred and take full responsibility.

・ 我們的確犯下了許多錯誤。

・ We've certainly made more than our share of mistakes.

・ 這一切都是無心之過。

・ Our mistake was never intentional.

・ 除了道歉之外，我不知道我們還能做些什麼。

・ Other than apologize, I don't know what else we can do.

・ 對於這件事我們真的很抱歉，你想要我們怎麼做？

・ We are sorry for the inconvenience—what else do you want?

・ 其實像這樣的小失誤到處都看得到。

・ This kind of mistake happens all time.

 ・ 每個人都會犯錯。

・ Nobody's perfect.

掌控發言權的
帝王學

操縱他人謀取自身利益的帝王學固然厚黑，
有些招數確實還蠻有效果的。但要同時運用
胡蘿蔔和棍子，就必須有心理準備，這套方
法絕不會讓你變成一位和藹可親的領導者，
卻能威脅利誘達到目的。

溝通哪些事

- 如何懷疑他人
- 如何威脅他人
- 從他人口中套出消息
- 反擊他人的指責
- 如何讓他人難堪
- 當有人插手管你的事
- 如何控制一場談話

- 如何和他人爭論
- 如何讓他人自我懷疑
- 避免討論拖延
- 如何迴避某人
- 如何營造信任

> 66
>
> *如果令人害怕和受人愛戴只能擇一，那麼令人害怕是較佳的選擇。*
> ——尼柯洛‧馬基維利
> （*Niccol Machiavelli*，十五世紀義大利政治思想家）
>
> 99

　　過去數十年來，「馬基維利」一詞已經成了操縱他人以謀取自身利益的形容詞。馬基維利式的領導者通常十分聰明且有遠見，但不見得有很高的道德操守。真正遵循馬基維利帝王學的人通常被認為是城府很深的算計者。當然，並不是每個優秀的領導人都採用這套帝王學，但是有些招數確實還蠻有效果的。要小心的是，這套學說絕不會讓你變成一位和藹可親的領導者。

 豎立防火牆

　　如果和你接觸的人對你懷著恐懼或覺得你捉摸不定，那麼你已經先佔了上風。不管其他人向你要求什麼，永遠表現出你還有更重要的事情要處理。

2 虛張聲勢

　　在你的周圍豎起一道牆隔絕大部分的人。限制其他人和你接觸的時間與機會。雖然這會讓對方憤怒，卻會讓你更容易操縱局勢。就像

帆船比賽一樣，你得把所有船壁上的堆積物清理乾淨才能達到最佳速度。另外，隔離他人也表示你不太在意他人的感受，十足帝王學。

❸ 威脅利誘

佛洛伊德曾說人為了兩個理由做事：獲得歡樂，或是避免痛苦。帝王學的信奉者知道如何運用胡蘿蔔和棍子來達到自己的目的。如果能成功威脅利誘，員工將做得更快更好，甚至更有自信。

❹ 遠離想沾你光的人

只要你能成功，不管你是怎麼辦到的，都會吸引一堆想沾光的人。除非有人引薦，否則不要多加接觸。如果某個人並非經由你信任的人介紹，突然出現在眼前，不要多加交談。和你面對面的機會越稀少，你的價值就越高。

112 如何懷疑他人

含蓄

- 你可以重講一次嗎？麻煩你。
- Would you repeat yourself, please?

- 很抱歉，你剛剛說什麼？
- I'm sorry—what did you say?

- 請稍等，我必須拿筆寫下來。
- Hold on, I need to write this down.

- 可以再請問一次你貴姓大名嗎？
- What is your name again?

- 你這麼說的理由是？
- What is your reason for saying that?

- 你怎麼會相信？
- Why would you believe that?

- 你為什麼這麼說？
- What makes you say that?

- 你到底想要表達什麼？
- What are you trying to say?

- 你怎麼會得出這樣的結論？
- How did you come to that conclusion?

- 我完全聽不懂你在說什麼。
- I'm not following you at all.

- 我真的很懷疑你怎麼會這樣想。
- I really wonder why you would think this.

- 你怎麼能這麼肯定？
- How can you be so sure?

- 看起來你也不是百分百肯定。
- You don't seem 100-percent sure of yourself.

- 你敢拿所有的身家財產來擔保嗎？
- Would you bet your life on it?

- 那只是你自己的說法吧！
- So you say.

- 那是你個人的想法。
- That's what you think.

- 嗯，這種說法很有趣…
- Interesting...

直接

- 隨便你…
- Whatever...

如何威脅他人

含蓄

- 我不知道現在我們還有什麼其他選擇。
  - I'm not sure where this leaves us.

- 你應該再考慮一下是不是真的要這麼做。
  - You might not want to do that.

- 我不認為這是個好主意。
  - I don't think that's the best idea.

- 你真的應該再想想這件事的做法。
  - You really should rethink your approach to this.

- 我不認為你應該照目前的方法進行下去。
  - I don't think it's healthy for you to continue on this track.

- 每個行動都會有不同的後果。
  - For every action, there is an equal and opposite reaction.

- 你應該了解每個行動都會帶來相對的後果吧？
  - You do know there are consequences to every action, right?

- 從現在開始，你要對自己的行為負責。
  - You are responsible for yourself from here on out.

- 我建議你不要這樣做。
  - I would advise you not to do that.

- 你會後悔的。
  - You're going to regret that

- 我警告你，每個行為都會帶來不同的後果。
  - I'm warning you, there are consequences to every action.

- 請你在做錯事／說錯話之前停止。
  - Please stop before you say/do something you'll really regret.

- 你最好先做最壞的打算。
  - You should prepare yourself for the worst.

- 你聽過因果報應嗎？
  - Have you ever heard of Karma?

- 總有一天會有人來制裁你的。
- Someone's going to cut you down to size one day.

- 你很快就會受到懲罰的。
- Your punishment will come in time.

- 惡有惡報，你知道的。
- Bad things happen to bad people, you know.

- 你會為此付出慘痛的代價。
- You're going to pay dearly for that.

- 你完蛋了！
- This is it for you!

明顯

- 走在路上你最好小心一點。
- You'd better not stop looking over your shoulder

114 從他人口中套出消息

含蓄

- 我保證，你說的一切都不會洩漏出去。
- I promise, anything you say will remain completely confidential.

- 我一個字也不會說。
- I am as silent as the grave.

- 你知道可以對我坦白的。
- You know you can tell me anything.

- 這件事就只有你我兩個人知道吧！
- This is just between the two of us.

- 我向你保證，這件事絕對不會傳出去。
- I give you my word that this won't go any further than these walls.

- 聽說你知道整件事的來龍去脈啊？
- Word on the street says you know what's going on.

- 你應該知道我們兩人之間的默契。
- You know that we have an understanding.

- 我只需要你簡單講一下事情的脈絡。
- I just need you to help me connect the dots.

- 我們可以把話攤開來講嗎？
- Can we cut to the chase?

- 能夠讓事情的主角來說明當然最好不過了。
- It would be nice to get it straight from the horse's mouth.

- 我想大概得和每個人都聊一聊，才能知道發生了什麼事。
- I guess I'll have to call around to get the facts.

- 你知道的，如果今天換了是我的話一定會全部告訴你的。
- You know I would tell you if the situation were reversed.

- 反正我一定會查清楚，不如你現在就告訴我吧！
- I'm going to find out anyway, so you might as well tell me.

- 你要自己告訴我，還是要我去問別人，看你囉？
- I can go behind your back or I can get it directly from you; your call.

- 難道你不相信我？
- Don't you trust me?

- 為什麼你對我的戒心會那麼重啊？
- Why are you holding back on me?

- 你這麼不信任我，真的讓我很傷心。
- I'm hurt that you wouldn't trust me with this information.

- 你到底在隱瞞什麼？
- What have you got to hide?

- 我問的每件事都不是什麼祕密啊！
- I'm not asking about anything that isn't common knowledge.

- 我可以整天都待在這裡喔！你還是趕快告訴我吧！
- I have all day to wait so why not just tell me?

- 沒有得到答案之前我絕對不會離開。
- I'm not leaving here without the information

- 如果你願意配合，事情會好辦很多。
- It would go much better for you if you just cooperated.

- 廢話少說，馬上給我我要的答案！
- Cut the crap and give me the answers I want!

強勢

- 或許你比較喜歡我用拳頭來問你？（玩笑口吻）
- Maybe you'd prefer that I beat it out of you? [joking]

115 反擊他人的指責

高雅

- 你說的根本就不是真的。
- What you say is simply not true.

- 我已經準備好面對並且澄清所有針對我的指責。
- I am prepared to defend myself against these allegations.

- 如果有人能證明我曾經做過這件事／說過這句話，請他站出來。
- If there is anyone who can categorically state that I did/said this, let him/her step forward.

- 我相信有很多人願意幫我做擔保。
- I know a dozen people who will vouch for me.

- 你這樣指控我，讓我很難過。
- I'm actually hurt that you would accuse me of something like that.

- 你不用白費力氣指責我，還不如把力氣拿去做更有效的運用。
- The energy you're using to blame me could be used in much more constructive ways.

- 這真的只是道聽途說啊！
- This is simply the result of rumor and innuendo.

- 看來你已經走投無路了，才會做出這種指控。
- These accusations are the work of a desperate person.

- 你真的覺得我會做這種事／說這種話嗎？
- Do you really think I would say/do something like that?

- 我為什麼要做那樣的事？
- Why would I do something like that?

- 我可以對天發誓：我沒有做這件事。
- I swear on a stack of Bibles that I didn't do it.

- 真的不是我。
- It wasn't me.

- 我是清白的。
- My conscience is clear.

- 我幹嘛要忍受這些指責！
- I don't deserve this.

- 在罵我之前，你應該先照照鏡子吧！
- Before you attack me, you should look in the mirror.

- 這些指控根本狗屁不通。
- These accusations are utter B.S.

- 你來罵我不過是想逃避別人對你的指控吧？
- Blaming me is the way you deflect blame from yourself.

- 你越是要抹黑我，就越顯得你自己有問題。
- The more you try to trap me, the more you point the finger at yourself.

- 你說的這些話代表你腦子真的有問題。
- Your statements make it clear that you are incompetent.

粗鄙

116 如何讓他人難堪

和善

- 請閉上你的嘴巴，到現在我的耳朵裡都還是你的聲音。（玩笑口吻）
  - Please turn your mouth off; I can still hear it running. [joking]

- 目前市場上混蛋缺貨，你還不快去報到。（玩笑口吻）
  - The jerk store called and they're running out of you. [joking]

- 我知道你盡力了。
  - Aw, you try so hard.

- 你真是可愛啊！
  - Aren't you cute.

- 好吧…（口氣猶豫）
  - Okaaaaay....[dubious]

- 真的嗎？
  - Really?

- 你怎麼老是做這種事啊？
  - Do you always do stuff like this?

- 哇，這真是個很「特別」的打扮／點子／想法。
  - Wow, that's an...interesting outfit/idea/thought.

- 我知道你想幫忙，其實我們自己就可以應付了。
  - I know you like to help, but I think we've got things under control.

- 看到你就讓我想起曾經年少無知的自己。
  - You remind me a lot of me when I was young and ignorant.

- 我不會和智商相差太多的人計較的。
  - I won't have a battle of wits with an unarmed opponent.

- 你自我感覺太過良好了吧？我的朋友。
  - You, my friend, are suffering from delusions of adequacy.

- 希望你今天已經準備妥當了。
  - I hope you're better prepared today.

- 門上的牌子寫得很清楚：只限專業人士！
  - The sign on the door reads, "Professionals only."

- 我希望至少能夠聽到一次有點智商的話。
- I'd love to hear something intelligent for a change.

- 需要你開口的時候我自然會叫你。
- When I need your opinion, I'll ask for it.

- 我根本不需要另外讓你難堪，你自己就行了。
- I don't have to make you feel stupid; you're halfway there on your own.

- 我欣賞你剛剛的發言，幸好還有幾句可以聽。（諷刺口吻）
- I liked what you just said very much—at least the parts that made sense. [sarcasm]

- 我真的很想對你好一點，只要你能再聰明一些些。（諷刺口吻）
- I'll try being nicer if you'll try being smarter. [sarcasm]

- 我需要你的程度，就好像小狗需要跳蚤一樣。（諷刺口吻）
- I need your help like a dog needs a tick. [sarcasm]

- 真抱歉！我的大腦會自動過濾廢話（諷刺口吻）
- I didn't hear you—my B.S. filter was on. [sarcasm]

- 到底要我說幾次，你才會離開啊？
- How many times do I have to flush before you go away?

嚴厲
- 你是怎麼活到現在的？
- How do you sleep at night?

117 當有人插手管你的事情

和善 · 三個和尚沒水喝。（玩笑口吻）	· Too many cooks spoil the broth. [joking]
· 嘿，我已經先佔領這個工作了。（玩笑口吻）	· Hey, I called dibs on this job first. [joking]
· 我已經處理得差不多了，但還是謝謝你。	· I've got it covered—thank you, though.
· 我想這個任務只需要一個人就夠了。	· I think this is a one-person operation.
· 謝謝，我想自己應付得來。	· Thanks, but I think I have a handle on it.
· 我已經在處理了。	· I have this covered already.
· 你應該已經做完自己的工作了吧？	· I assume you've already done the jobs you were assigned?
· 或許你可以去幫忙其他的事情。	· Maybe there is something else you can do.
· 可以請你給我一點做事的空間嗎？	· A little breathing room, please?
· 你應該把注意力放在自己的事情上。	· Maybe you should worry about your own stuff.
· 請不要介入和你沒有關係的事情。	· Please don't get involved in things that don't concern you.
· 你沒有其他的事好做了嗎？／你沒有其他地方好去了嗎？	· Don't you have something else you need to do/somewhere else you need to be?
· 你沒有更需要你的事情可以做嗎？	· Don't you have anything better to do?

- 為什麼不先管好你自己的事情呢？
- Why don't you take care of your side of the street instead?

- 可以請你還給我一個清淨的空間嗎？
- Would you please grace this space with your absence?

- 我自己就能處理了，但還是謝謝。（諷刺口吻）
- I've got things under control, but thanks. [sarcasm]

- 有人請你加入了嗎？
- Um, who invited you to this party?

嚴厲
- 給我離開。
- Get lost.

118 如何控制一場談話

禮貌
- 請容許我說一下…
- If you would allow me to speak for a moment...

- 或許你不太了解我的意思。
- Perhaps you didn't understand me.

- 我可以說句話嗎？
- May I just say this?

- 抱歉，我還沒講完。
- Excuse me, I'm not finished.

- 我的話還沒有結束。
- I haven't finished what I was saying.

- 不好意思！
- Pardon me!

- 你真的沒有必要再繼續說下去了。
- There's really no need for you to go on.

- 我想今天到此為止吧！
- I think I've heard enough.

- 這件事就此結案。
- This case is closed.

- 大家不要再歹戲拖棚了。
- Someone needs to stop beating this dead horse.

- 這件事已經沒有繼續討論的必要。
- There simply isn't anything more to say on the matter.

- 你的論述根本沒有意義。
- Your reasoning is going nowhere.

- 你只是在浪費時間、精力而已。
- You're wasting your time and energy.

- 你這是什麼態度啊？
- What kind of mentality is that?

- 你的父母沒教過你要尊重別人嗎？
- Didn't your family teach you respect?

- 我想你說的已經夠多了。
- I think you've said enough already.

- 我很想繼續聽你閒扯下去，但是我還有其他重要的事要處理。
- I'd be more than happy to let you ramble on, but I've got more important things to do.

- 你已經說得夠多了，現在換你聽我說。
- I'm done listening to you; now it's your turn to hear me out.

- 你是不懂「到此為止」的意思嗎？（諷刺口吻）
- What part of "done" don't you understand? [sarcasm]

- 無意冒犯，但是可以請你閉嘴嗎？
- No offense, but would you please just shut up?

粗魯

- 順便告訴你，這場談話現在開始由我主導。
- Just FYI: I will be controlling this conversation from now on.

如何和他人爭論

含蓄

- 是沒錯，但是…
- Yes, but...

- 我同意，可是…
- I agree, but...

- 你可能是對的，不過…
- You may be right, but...

- 那只是你的看法。
- So you say.

- 隨便。（視情況重複使用）
- Okay, whatever. [repeat as necessary]

- 在結束之前，我想再提出一件事情。
- Before we finish, let me just say one last thing.

- 我要提出最後一點，然後就散會。
- Let me just say one more thing and then we're done.

- 好吧！但是我不會就此放棄。
- Okay, but I'm not letting this go.

- 我會另外找個時間繼續討論這件事。
- We'll continue this at a later date of my choosing.

- 我晚點／明天／另外找時間回來處理這件事。
- We'll talk more about this later/tomorrow/some other time.

- 我稍後／明天／下週會處理這件事。
- I will follow up about this later/tomorrow/next week.

- 不，今天就到此為止。
- No, I'm done here for now.

- 好吧，我先讓你離開。
- Okay, I guess I'd better let you go.

- 以上就是我對所有問題的看法。
- Here is my summation of the issue as I see it.

- 我們來總結吧！已經討論得差不多了。
- Let's wrap it up; there's really nothing more to say.

- 你大概講到了事情的重點，但那並不是結論。
- That may be the long and the short of it, but it's not the end of it.

- 你已經說得很多了，謝謝。
- You've talked enough, thank you.

- 你只是重複在講同樣的事情。
- Now you're just repeating yourself.

- 我不會改變我的決定，所以你不用再多費唇舌了。
- I'm not going to change my mind, so it's best that you stop talking.

- 我只想再講最後一件事，然後結束討論。
- This is the last thing I want to say and then this conversation is finished.

直接

120 如何讓他人自我懷疑

含蓄

- 我知道你是一番好意。
- I know you mean well.

- 我相信你已經盡力了。
- I'm sure you're doing the best you can.

- 畢竟你並不知道這些事。
- How could you have known?

- 嗯，我聽到的不是這樣喔！
- Hmm, that's not what I heard.

- 嗯，你可能要再查證一下資料來源。
- Hmm, you might want to consider the source.

- 這和我聽到的情況完全不一樣。
- I was led to believe something quite different.

- 某某某和我說的不是這樣。
- That's not what so-and-so told me.

- 你從哪裡得到這些資料的？
- Where did you get your information?

- 我很好奇，你究竟在想些什麼啊？
- I'm curious—what were you thinking?

- 你為什麼要做／說這種事情呢？
  - Why would you do/say something like that?

- 難道你對這件事不生氣／不難過嗎？
  - Aren't you angry/upset about that?

- 如果我是你，一定氣炸了。
  - If I were you, I'd be pretty P.O.'d.

- 有時候我真的搞不懂你。
  - Sometimes I just don't get you.

- 你是說你並不知道…嗎？
  - You mean you didn't know that...?

- 你竟然不知道這件事情，實在太遺憾了。
  - It's a shame you didn't know about this.

- 通常你的想法都寫在臉上。
  - It's usually pretty obvious what you're thinking.

- 老實說你的態度／行為讓我有點吃驚。
  - Honestly I'm a bit surprised by your attitude/actions.

- 如果我是你，一定會非常憤怒／難過／受傷。
  - If I were you, I would be so mad/upset/sad/hurt.

- 你的想法和大家都不一樣。
  - Nobody else I know thinks that.

- 你怎麼能夠忍受這些呢？
  - How can you tolerate that?

- 你的表現完全和我想的不一樣。
  - This is not really what I expected from you.

- 我真的不懂你為什麼會有那種反應。
  - I'm really not sure why you're reacting that way.

- 如果是我就不會這樣做，不過算了。
  - That's not how I would have done it, but whatever.

- 我想你應該知道什麼是最好的…（猶豫口吻）
  - I'm sure you know what's best.... [dubious]

消極

121 避免討論拖延

專業

- 我們先把這個問題放一邊，繼續討論其他更要緊的事情。
  - Let's set aside this topic and move on to more pressing matters.

- 這件事可能過幾天再來談會比較好。
  - This topic of conversation might be better suited for another day.

- 我們可以過陣子再來討論這件事嗎？
  - Can we put this off until a later date?

- 我覺得我們應該另外找時間討論這件事。
  - I think we should discuss this some other time.

- 這是個複雜的問題，我們先略過不談。
  - This is a touchy subject—let's steer clear of it.

- 這是個敏感的議題，最好是能避免討論。
  - This is a controversial subject— probably best to avoid it.

- 這只是我個人的想法，但我認為這個問題不需要討論。
  - I don't think we need to address that, but that's just my humble opinion.

- 抱歉，我現在不想談論這件事情。
  - I'm not comfortable talking about this right now, sorry.

- 我認為我們不需要討論這件事情。
  - I'm pretty sure we don't need to get into this

- 建議就把這個議題從會議流程中刪掉吧！
  - Perhaps we can just cross that off the agenda.

- 這個討論可能火藥味會太重。
  - There is a bit too much controversy surrounding this conversation.

- 我不想繼續討論這個議題了。
  - That is a subject I don't wish to continue discussing.

- 這個問題已經超過我想討論的範圍。
  - That subject is off limits in my book.

・或許我們不應該討論這麼敏感的話題。	・Maybe we shouldn't talk about such a touchy subject.
・再討論下去也只是浪費時間。	・It would be a waste of our time to discuss this any further.
・我們何必要討論這件事呢？	・Why talk about this if we can avoid it?
・我不認為這件事有這麼重要。	・I don't believe that this is important.
・我們有必要繼續談論這件事情嗎？	・Is this really important enough to continue discussing?
・對這件事最好的處理方法就是不要理它。	・The best way to talk about this is to not mention it at all.
・我是可以講這件事，但何必浪費其他人的時間呢？	・I could talk to you about this, but why waste everyone's time?
・你要知道，這個話題非常敏感的。	・I hope you realize that this topic is a sensitive one.
・你現在講到了一個很麻煩的問題。	・You're getting into a big subject, here.
・或許我們應該把這個討論擱一邊，永遠擱一邊。	・Maybe we should just table this discussion...forever.
・我不討論這個問題。	・This topic is not open for discussion.
・繼續討論這件事一點意義也沒有。	・We'd just be beating a dead horse talking about this.
・我暫時不想討論這件事。	・This discussion will not be happening anytime soon.
・抱歉，但我不打算討論那件事情。	・Sorry, but I won't discuss that.
・這件事以後不要再提了。	・There will be no further deliberation concerning this.

- 這件事已經沒有任何討論的餘地了。
- This is not open for discussion or debate.

- 我不會再做任何回應了。
- I'm not going to entertain that notion.

- 我不想再聽到任何相關的字句。
- I'm simply not going to listen to another word.

- 我一個字也不想聽。
- I don't want to hear a word about it.

衝動
- 不要再說了，好嗎？
- Just drop it, okay?

122 如何迴避某人

圓滑
- 哇，今天大家居然都搶著要我耶！（玩笑口吻）
- Wow, everyone wants a piece of me today! [joking]

- 很抱歉，但我現在真的是分身乏術。
- I'm sorry, but I've got too many plates in the air at the moment.

- 我很想和你聊聊，但是今天真的是抽不出時間。
- I'd love to talk to you, but I haven't got a minute today.

- 我也想待久一點，無奈有任務在身。
- I wish I could linger, but duty calls.

- 我會盡快找時間和你聚聚，我保證。
- I'll try to catch up to you some other time, I promise.

- 我很想和你多聊聊，但正好有事要出門。
- I'd like to chat, but I'm just running out the door.

- 你要不要先和我的秘書預約時間啊？
- If you want, you can make an appointment with my secretary.

- 我真的很忙，你可以發個電子郵件／留個話／和我的助理聯絡嗎？
- I'm so busy—can you shoot me an email/leave a voicemail/talk to my assistant?

- 我真的很想和你聚一聚，但剛剛好要出門。
- I'd love to hang out, but I'm just on my way out.

- 或許我們可以另外找個時間討論？
- Perhaps we can discuss this some other time?

- 很遺憾，我現在真的沒時間處理這件事。
- I just don't have the time right now, unfortunately.

- 如果可以，我真的不想討論這件事。
- I'd rather not discuss that, if that's okay with you.

- 我現在不想講這件事。
- I don't want to talk about it now.

- 我們不是已經講過了嗎？
- We already talked about this, didn't we?

- 這真的有討論的必要嗎？
- Is this conversation really necessary?

- 何必沒話找話說呢？
- Why stir the pot when we can call it a day?

- 繼續講下去也沒什麼好處。
- Continuing to hash this out isn't productive.

- 如果你還想講下去，請直接留言在我的語音信箱。
- If you still feel the need to hash it out, please leave a message on my phone.

- 我再也不想和你討論了。
- I don't wish to ever have that conversation.

- 我不想再說了。
- I don't wish to deliberate any further.

- 我們有討論的必要嗎？
- Why should we discuss anything?

- 抱歉，我真的沒興趣。
- Sorry, I'm really not interested.

・對不起，就當這段對話沒發 生過。
・Excuse me, but this conversation is not happening.

・為什麼我要和你討論這件事 啊？（嘲諷口吻）
・Why would I discuss this with you? [sarcasm]

・很抱歉！我想這個討論永遠 不會發生。
・This discussion will never happen— sorry!

・我絕對不會和你討論這件事 情的。
・You're the last person on earth I would discuss this with/hang out with.

・唉呀！時間都已經這麼晚了 啊！
・Oops, look at the time!

無禮 ・休息一下吧！
・Take a hike!

123 如何營造信任

積極 ・我永遠不會讓你失望的。
・I will never let you down.

・我說話向來一言九鼎。
・My word is my bond.

・我願意為你赴湯蹈火。
・I would walk through fire for you.

・我一旦承諾／答應，就一定 會辦到。
・When I make a commitment/promise, I always keep it.

・對這件事我很有信心。
・I have a high degree of confidence regarding the situation.

・我是個誠實正直的人。
・My integrity is unimpeachable.

・我有很好的預感。
・I feel great about this.

・一切都已經準備就緒。
・All systems are go.

- 請你相信我，我不會讓你失望的。
- Please put your trust in me—I won't let you down.

- 無論如何我一定會完成任務。
- I will stop at nothing to make this happen.

- 相信我，事情會很順利的。
- Trust me, it will be taken care of.

- 我向你保證，你可以信任我。
- You can trust me, I promise.

- 對我來說，名聲勝於一切。
- My reputation precedes me.

- 請相信我就像我相信你一樣。
- Believe in me just as I believe in you.

- 大家都知道的，我一向不負所託。
- I am known for my trustworthiness.

- 我在〔某主題〕上有多年的經驗。
- I have many years of experience in [topic at hand].

- 我做人一向以誠信為本，我保證。
- I am a person of integrity, I assure you.

- 我一向是實話實說。
- I tell the truth in all matters.

- 你可以到處問，我的名聲很好的。
- Ask around. I have a good reputation.

- 通常當其他人都放棄的時候，我還是會堅持下去。
- I've kept going when others might have stopped.

- 我會盡一切努力讓事情成功。
- I will make every effort to make this happen.

- 對於結果我很有信心。
- I feel at peace about the outcome.

- 如果事情出了任何差錯，我會負全部責任。
- I'll hold myself responsible if things go awry.

- 你有任何理由不相信我嗎？
- You have no reason to distrust me.

- 我沒有理由要騙你，對吧？
- I have no reason to lie to you, do I?

輕鬆

Bonus

領導者可以這樣說

任何人碰上預料之外的狀況都會感到棘手，
特別是在上位的領導者。雖然這一點也不輕
鬆，如果你嘴拙，說話不懂得修飾，那壓力
更是大。提供你一些點石成金的詞句，協助
你成為大方得體的明日之星。

溝通哪些事

- 領導者如何表同意
- 領導者如何說不
- 領導者如何說或許
- 領導者如何說不知道
- 領導者如何說我不懂
- 領導者如何說要保密

124 領導者如何表同意

肯定

・理所當然！	・Absolutely.
・我很肯定。	・Positively.
・這是當然！	・Certainly.
・我同意！	・I agree.
・我贊成！	・I concur.
・當然！	・Of course.
・完全正確！	・Precisely.
・一點也沒錯！	・Exactly.
・毫無疑問！	・Without a doubt.
・非常同意！	・Very much so.
・這是一定的！	・That's for sure.
・不會有問題的。	・It's a sure thing.
・是這樣沒錯！	・That's exactly the case.
・我也是這麼想。	・I'm of the same opinion.
・一句話，是的！	・In a word, yes.
・我們的想法一致。	・We're of the same mindset.
・我們站在同一陣線上喔！	・We're on the same wavelength.
・這件事情上我完全同意你。	・We're of one accord on that.
・你完全看穿了我的心。	・Hey, you're reading my mind.

- 這件事我站在你這邊。 · I'm with you on that.

- 我把你的話聽進去了。 · Message received.

- 如果一定要表態，那麼我同意！ · If I have to give an answer, then it's yes.

猶豫 · 我想應該是這樣的。 · I'm pretty sure that's the case.

125 領導者如何說不

肯定 · 絕對不可能！ · Under no circumstances.

· 這件事沒有商量的餘地。 · That's out of the question.

· 不可能！ · That's impossible.

· 絕對不行！ · Absolutely not.

· 抱歉，就是不行！ · I'm sorry, but no.

· 不行！ · No can do.

· 我不能同意。 · I can't agree to that.

· 我必須要拒絕。 · I'm afraid that's not an option.

· 我必須告訴你我反對。 · I have to humbly disagree.

· 我尊重你的意見，但是無法同意。 · I'm going to have to respectfully decline.

· 我不會那麼說。 · I wouldn't necessarily say that.

· 我不這麼認為。 · I don't get that impression

- 我並不是這樣想的。
- Not to my knowledge.

- 在我看來事情不是這樣的。
- It doesn't seem that way to me.

- 我認為這並不合理。
- It doesn't make sense to me.

- 我想我應該會反對。
- I tend to disagree.

- 我不認為是這樣。
- I don't think so.

- 我目前不打算這麼做。
- That wouldn't be my preference right now.

猶豫
- 事情不應該這樣，當然我也有可能是錯的。
- I don't think that's correct, but I could be wrong

126 領導者如何說或許

禮貌
- 有可能！
- Perhaps.

- 我們可以再討論。
- That's debatable.

- 事情應該還有轉圜的餘地。
- It's not implausible.

- 事情應該還有其他可能性。
- It's not outside the realm of possibilities.

- 這件事還有討論的空間。
- It's completely open for discussion.

- 我並不確定。
- I'm not sure.

- 我還有些疑問。
- I'm not convinced.

- 事情還有轉機的。
- It's not out of the question.

- 我現在還不能肯定。
- I couldn't say for sure.

我無法確定。	I'm not sure on the facts.
還有很多需要思考的地方。	It is worth reflecting upon.
兩方面都有可能。	I could go either way.
非常有可能。	It's possible.
目前情況還不太明朗。	That's iffy
一切有待證實。	That's questionable.

隨便 · 天知道！ · Who knows?

127 領導者如何說不知道

肯定

我完全不知情。	I have absolutely no idea.
抱歉，我真的不知道。	I'm sorry, but I simply don't know.
相信我，我也很想知道。	Believe me, I wish I knew.
這件事情我毫無頭緒！	I haven't a clue.
很多事實我都不知道。	I don't have all the facts to give a qualified response.
我對事情的了解還不夠，無法做出任何結論。	I'm not informed enough to give a precise response.
我需要了解更多資訊才有辦法做出結論。	I'm not knowledgeable enough to have an opinion on the matter.
我不知道，但是我會找出答案的。	I don't know, but I'll find out.

- 別急，我會找出答案的。
- Hang on, I'll look it up.

- 這個問題很好，但是我目前沒有任何答案。
- That's a good question, but I don't yet know the answer.

- 我還沒辦法給出任何肯定的答案。
- I can't say anything definitive either way.

- 我需要收集更多的資訊。
- I will need to do more research.

- 我必須承認自己的無知。
- I'm going to have to plead ignorance.

- 在這件事情上，我掌握的資訊還不太夠。
- I don't have enough information on the subject.

- 我不確定，但可以幫你問問。
- I'm not sure—let me find out for you.

- 我真的不知道啊！
- I can't really tell.

- 我沒辦法給你確定的答覆。
- I'm not exactly sure.

- 我可以給你答案，但可能是錯的。
- I could give an answer, but it wouldn't necessarily be correct.

- 我沒有好答案可以給你。
- I don't have a good answer for that.

- 我還不知道要如何回應。
- I have no response at this time.

- 我根本沒聽過這些東西，要怎麼回答啊？
- How can I reply if I don't know what I'm talking about?

猶豫

- 我不這麼認為，但也無法完全肯定。
- I don't think so, but I am not entirely sure.

128 領導者如何說我不懂

優雅

- 抱歉，我不是很明白你的意思。
  - I'm sorry, but I'm not sure that I understand you entirely.

- 可以麻煩再解釋一次嗎？
  - Would you be so kind as to explain that to me once more?

- 真不好意思，我剛剛沒聽清楚你的意思。
  - I'm sorry to say this, but I'm not following you.

- 我不確定是不是完全了解你的意思。
  - I'm not certain if I understand you properly

- 請再說明一次。
  - Please help me by clarifying.

- 抱歉，我沒辦法一口氣消化那麼多資訊。（玩笑口吻）
  - Sorry, my mind can't process information that quickly. [joking]

- 我不是很明白你說這些話的意思。
  - I am not sure if I understand where you're going with this.

- 我剛剛可能漏掉了幾個重點。
  - I'm probably missing something, here.

- 看來我得多做點功課了。
  - I must read up on the subject.

- 我不知道你想說的到底是什麼。
  - I'm not sure what you're trying to say.

- 你能不能用外行人聽得懂的話再說一次？
  - Can you say that again in layperson's terms?

- 我是不是錯過了什麼？
  - Am I missing something?

- 我聽不懂。
  - I don't get it.

- 我完全不懂你在說什麼。
  - I have no clue what you're talking about.

- 你把我搞糊塗了。
  - Well, I'm lost.

粗魯

- 到底怎麼一回事？
  - What the heck?

129 領導者如何說要保密

優雅

- 抱歉，我不能針對這件事情發言。
- I'm sorry, but I'm not at liberty to talk about that with you.

- 抱歉，我不能提供任何關於這件事的資訊。
- I'm sorry, but I can't disclose any information on that topic.

- 你真的想知道不妨試試用讀心術吧！（玩笑口吻）
- You'd have to read my mind to find out. [joking]

- 我發過毒誓要保守祕密的。（玩笑口吻）
- I'm under oath to keep silent on that. [joking]

- 如果我說了，那就要殺你滅口囉！（玩笑口吻）
- If I told you I'd have to kill you. [joking]

- 這是國家的最高機密！（玩笑口吻）
- That info is classified as Top Secret. [joking]

- 抱歉，我無法透露任何相關的訊息。
- Sorry, I am not allowed to disclose that type of information.

- 可惜我無法公開發表任何的評論。
- Unfortunately there's no way I can speak openly about this.

- 我有保持沈默的權利吧！
- I have to plead the Fifth on that.

- 我必須保守祕密。
- It was told to me in confidence.

- 我不會洩露任何機密的。
- I never break a confidence.

- 我不能越俎代庖。
- I don't want to step on any toes, here.

- 這個問題不應該由我來回答吧！
- I'm not the best person to answer that question.

- 我沒有資格提供任何答案。
- I'm not in a position to give you a complete answer.

- 任何的發言都是不恰當的。
- I would be speaking out of turn if I said anything.

- 有些事還是閉口不談的好。
- Some things are better left unsaid.

- 我不認為這件事應該被提出來討論。
- I don't think it's appropriate to talk about this.

- 為什麼你要知道這件事呢？
- Why do you need this information?

- 你應該知道我不能說的。
- You know very well I can't talk about that.

- 別插手這件事。
- Leave it alone.

- 這是人家的私事！
- It's under wraps.

- 這不關你的事吧！
- It's none of your business.

- 你已經越界了喔！
- You're nosing around where you shouldn't be.

- 你別管！
- Keep out of it.

粗魯
- 我為什麼要告訴你？
- Why would I talk to you about this?

最有分寸溝通術【長銷經典版】：
即使「你滾蛋！」也能說得漂亮，成為職場人氣王（老中老外都能通）
The Leader Phrase Book: 3000+ Powerful Phrases That Put You In Command

作　　　　者　派屈克‧亞倫（Patrick Alain）
譯　　　　者　陳松筠
特 約 編 輯　吳欣恬
封 面 設 計　巫麗雪
內 頁 排 版　陳姿秀
行 銷 企 劃　林瑀、陳慧敏
行 銷 統 籌　駱漢琦
業 務 發 行　邱紹溢
營 運 顧 問　郭其彬
責 任 編 輯　賴靜儀
總 編 輯　李亞南
出　　　　版　漫遊者文化事業股份有限公司
地　　　　址　台北市松山區復興北路 331 號 4 樓
電　　　　話　（02）27152022
傳　　　　真　（02）27152021

讀者服務信箱　service@azothbooks.com
漫遊者書目　　www.azothbooks.com
漫遊者臉書　　www.facebook.com/azothbooks.read
劃撥帳號　　　50022001
劃撥戶名　　　漫遊者文化事業股份有限公司
發　　　　行　大雁文化事業股份有限公司
地　　　　址　台北市松山區復興北路 333 號 11 樓之 4

三版一刷　　　2022 年 4 月
定　　　　價　台幣 420 元
Ｉ Ｓ Ｂ Ｎ　978-986-94362-3-6
版權所有‧翻印必究（Printed in Taiwan）
◎本書如有缺頁、破損、裝訂錯誤，請寄回本公司更換。

漫遊，一種新的路上觀察學
www.azothbooks.com

漫遊者

漫遊者文化

大人的素養課，通往自由學習之路
www.ontheroad.today
遍路文化
on
the road
遍路文化‧線上課程

國家圖書館出版品預行編目 (CIP) 資料

最有分寸溝通術【長銷經典版】：即使「你滾蛋！」也能
說得漂亮，成為職場人氣王（老中老外都能通）/ 派屈
克．亞倫（Patrick Alain）著；陳松筠譯. – 三版. – 臺北市
：漫遊者文化出版：大雁文化發行, 2022.04
304 面；17×23 公分
譯自：The leader phrase book : 3,000+ powerful phrases
that put you in command
ISBN 978-986-489-623-3（平裝）

1. 商務傳播 2. 職場成功法

494.2   111004992